OGALLALA
BLUE

OGALLALA BLUE

WATER AND LIFE

ON THE HIGH PLAINS

WILLIAM ASHWORTH

The Countryman Press
Woodstock, Vermont

Copyright © 2006 by William Ashworth

For information about permission to reproduce selections from this book,
write to Permissions, The Countryman Press, P.O. Box 748, Woodstock, VT
05091

Cover photo © Lee Rentz
Cover design by Karen Schober
Interior design by Chris Welch

Library of Congress Cataloging-in-Publication Data
Ashworth, William, 1942–
Ogallala blue : water and life on the High Plains /
William Ashworth. — 1st pbk. ed.
p. cm.
ISBN 978-0-88150-736-2 (alk. paper)
1. Irrigation water—High Plains (U.S.)—History. 2. Ogallala Aquifer—
History. 3. Irrigation—High Plains (U.S.)—History. 4. Agriculture—High
Plains (U.S.)—History. 5. Agricultural ecology—High Plains (U.S.)—
History. I. Title.
S616.U6A78 2007
553.7'90978—dc22
2007014447

Published by The Countryman Press,
P.O. Box 748, Woodstock, VT 05091

Distributed by W. W. Norton & Company, Inc.,
500 Fifth Avenue, New York, NY 10110

10 9 8 7 6 5 4 3 2 1

For Melody,
who is present on every one of these pages

Civilization has been a permanent dialogue between
human beings and water.

— Paolo Lugari,
founder of the Gaviotas Community in Colombia

In their efforts to provide a sufficiency of water where there was
not one, men have resorted to every expedient from prayer to dynamite.
The story of their efforts is, on the whole, one of pathos and tragedy, of a
few successes and many failures.

— Walter Prescott Webb, *The Great Plains*

What'cha gonna do when the well runs dry?
Gonna sit on the bank, watch the crawdads die.
Honey, oh sugar baby of mine.

— American folk song

CONTENTS

CONTENTS

THREE: MINIMUM WATER

An *acre-foot* is the amount of water necessary to cover one acre of land one foot deep. It is equivalent to 43,560 cubic feet, or 325,851 gallons.

OGALLALA
BLUE

THE PALISADED PLAIN

GAUNT CORONADO, chasing fabled Quivira, marched his ironclad army east and south along the stream he called El Rio Cicúique, guided by a taciturn native known only as "the Turk." Throwing a rude bridge across the river—today's Pecos—near the present-day town of Santa Rosa, New Mexico, Coronado and his men trailed across. On the far side rose the Mescalero Escarpment, a band of tall cliffs "like a palisade," bending far out of sight both north and south. This was in the spring of 1541, and the weather had already turned hot. The Spaniards in their metal helmets and chest plates sweated up a steep break in the palisade, leading their skittish horses. At the top they paused, blinking. The corrugated country they had been traveling through ended. Before them sprawled an immense plain, larger than Portugal, featureless, and flat as an all-encompassing sea.

It is impossible, today, to reconstruct with any certainty the next part of the expedition's journey. The Turk was intentionally misdirecting them, leading them away from the pueblos they had despoiled near what are now Albuquerque and Santa Fe, trying to get them lost. The tale of Quivira, where people ate from plates of pure gold and the king in his finery was lulled to sleep by a symphony of little gold bells, was the bait; the trap was this uniform immensity of flatness. Coronado and his men fell into it. Soon they were wandering aimlessly, all sense of direction gone, over an endless surface of short, gray-green grass that left no trace of their passage. The sky was a blue bowl resting upon blue horizons. They saw vast herds of shaggy brown "cows"—the first European encounter with the American buffalo. Disorientation was constant. There were, wrote Coronado, "no more landmarks than if we had been swallowed up by the ocean . . . not a stone, nor a bit of rising ground, nor a tree, nor a shrub, nor anything to go by." Sorties sent out for game wandered in confused circles. At least one party never found its way back.

The worst thing was the lack of water. This huge, level landscape, though grassy, was nearly bone-dry. The few small lakes were shallow and saline; the few emaciated streams were warm as baths, slimey and green with buffalo dung. The Spaniards were constantly thirsty and constantly sick. Finally, a month or more after entering this Plain of the Palisade, this *Llano Estacado*, they stumbled out at its eastern edge and descended into a red *barranca*, identified by modern historians as Blanco Canyon, near Crosbyton, Texas. Here, Coronado—who must have become suspicious of his "guide" long before—encountered a traveler who identified himself as a member of the Quivira tribe, a people who later became known as the Wichita. No, no, the traveler protested, they were going the wrong way. Quivira was to the north, not to the east. Far, far to the north.

Dragging the Turk along—he was now effectively a prisoner—the expedition heaved itself around and headed north "by the needle," across league upon league of flat green prairie interrupted occasionally by shallow, eastward-flowing streams: the Canadian, the Beaver, the Cimarron, the Arkansas. Eventually, near what is now Lyons, Kansas, they came to the country called Quivira. There were no gold plates and no gold bells, only buffalo-robed people in buffalo-skin huts. Coronado had the Turk brought before him and strangled. He allowed his troops three weeks' rest, then set a course back toward the

Pecos—having just missed the true lode of treasure that lies buried beneath North America's dry heart. Though the would-be conquistador could not know it, this treasure had been close at hand during almost all of the journey from the Mescalero Escarpment to Quivira and back again. While his thirst-maddened men drank from buffalo wallows, billions of gallons of fresh, cool water lay, well concealed, only a few yards away.

ONE

THE
UNDERGROUND
OCEAN

I

THE PHANTOM RESERVOIR

O NE DOES NOT GENERALLY lose an Army Corps of Engineers' dam, so it is notably odd when you can't find one. I had left the tidy little city of Guymon, Oklahoma, early on a cool spring Sunday, headed east toward a large, blue, reservoir-shaped blob athwart the Beaver River that the people at Rand McNally had labeled Optima Lake. The sun struggled through a high, thin layer of haze; the flat green surface of the Oklahoma Panhandle glittered with myriad tiny rainbows, one in each droplet of water shed by the dozens of slowly moving center-pivot sprinklers. Somewhere along US 412 I should have found signs directing me to the dam. Where were they?

Acting on a hunch, I turned north onto a paved side road marked by a small hand-lettered sign that offered "camping." Two miles up the road I came to a much larger sign, a curvilinear shape of decrepit, brown-painted wood mounted on a weed-covered base made from cemented chunks of sandstone: standard-issue Corps of Engineers recreation area construction, but badly gone to seed. The sign had originally said WELCOME TO OPTIMA DAM. Those words had been covered by a big blue placard with white lettering, which read:

> **Attention Visitors**
>
> This park is closed, although
> access will remain open. No
> services are provided. Please
> do not litter. To report area
> abuse call 405–766–2701.

It was another three quarters of a mile down a branch road to a deserted parking lot at the edge of a bluff overlooking the confluence of the Beaver River and Coldwater Creek. Grasses and yuccas grew up through checkerboard cracks in the asphalt. An imposing restroom facility was locked and deteriorating into shabby irrelevance. A faded sign attached to the remnants of a cyclone fence warned anglers that any bass caught were to be returned to the lake immediately. There was no lake. The dam's immense earthen pile, 3 miles long and 120 feet high, loomed to the east, its square concrete intake structure towering forlornly beside it. Two little riverlets trickled in and pooled in a depression in the valley floor. The pond they made there failed to reach even to the base of the dam.

Ever since its completion in 1978 at a cost of forty-six million dollars, Optima Dam has looked exactly like this: a huge pile of expensive dirt in front of a puddle of water. The intake structure has so far taken in nothing but air; the carefully wrought upstream slope of the dam, made to withstand the pressure of several cubic miles of lake water, has withstood nothing damper or more stressful than raindrops. Designers of dams specify three reservoir depths, called *maximum, minimum,* and *conservation pools.* The last-named is the shallowest level the reservoir is expected to drop to in the driest years. In the twenty-five years since its completion, Optima Lake has never even approached the conservation-pool level.

Where is the water that should be filling the phantom reservoir behind Optima Dam? Where is the water that should be running down Running Water Draw in Texas, or springing forth from Wagon Bed Spring in Kansas, or welling from the earth at Cheyenne Wells, Colorado? What is wringing the dry center of North America even drier?

Look in your cupboard. Look in your refrigerator. Look in your closet. The water is there—pulled from beneath the plains and shipped to a shopping center near you, as food, as drink, and as clothing.

WATER IS LIFE. It is our primary support system, the chief component of our tissues, and the only substance that all living things must have or die. There are bacteria that can live without oxygen; there are cave creatures and deep-sea dwellers that can live without sunlight. Nothing can live without water. Water grows our food, floats our boats, flushes our waste, builds our bodies, and pumps through us—thinly disguised as blood—at roughly one heartbeat per second. Civilizations rise and fall on it. The great dams of the United States are the hallmark of our culture as much as the flood irrigation systems of Egypt, the levees of China, the aqueducts of the Roman Empire, and the canals of Mesopotamia were hallmarks of theirs.

Those who live in humid lands can afford to take water for granted. Arid-land dwellers know better. Where little rain falls, each drop is a benediction. A glass of water is a miracle; a crop is an engineering feat. Rainwater must be captured, tamed, transported, hoarded for an unrainy day. Without irrigation, an arid-land farm cannot exist.

The High Plains region of North America is officially classed as semiarid, not arid, but the distinction does not matter much. Little rain falls on the wide earth of this immense, dun-colored expanse once roamed by the buffalo peoples, the Cheyenne and the Pawnee and the Oglala and the Comanche. The Rockies wring moisture from the western sky; winds blowing north from the Gulf of Mexico swing to the right, dropping their wet gifts well to the east of the hundredth meridian. Thunderstorms and tornados wreak brief havoc, and the southward swing of the jet stream in winter brings cold storms down from Canada, but for most of the year, over most of the plains, the sky is Ogallala blue.

The explorers and surveyors who followed Coronado into this land called it the Great American Desert. They thought it worthless for farming; today, it is one of the prime agricultural areas of the world. The magic that has made this possible is another Ogallala blue—a wide, dark blue that hides beneath the soil. Groundwater is the glass slipper that has transformed this Cinderella landscape into a princess. Under the sand hills, under the shortgrass prairie, under the rich harvest of corn and wheat and cotton, lurks an ocean: the Ogallala Aquifer. It sprawls from central Texas to southern South Dakota and from eastern Colorado almost into Iowa, and there is enough water in it to fill Lake Erie. Nine times.

It is hard to overestimate the impact that this bounty of buried water has had on American life. If you snack on popcorn or peanuts, you are probably eating Ogallala water; if you dress in cotton clothing, you are probably wearing it. The Ogallala grows wheat and milo, sunflowers and sorghum. It grows alfalfa for cattle, and it grows the cattle as well. It provides drinking water for large parts of eight states. The fourteen million acres of crops spread across its flat surface account for at least one-fifth of the total annual U.S. agricultural harvest. Five *trillion* gallons of water are drawn from the Ogallala annually—about 30 percent of all groundwater used for irrigation in the United States. If the aquifer went dry, more than $20 billion worth of food and fiber would disappear immediately from the world's markets.

Or perhaps we should say *when* the aquifer goes dry. That is not hypothetical doomspeak; it is happening. Not overnight, not next week, but steadily, stealthily, and for the most part irreversibly. The bulk of the water here is what geologists call fossil water—dampness from a distant era, preserved in earth and stone like the bones of dinosaurs. Most of it arrived during the aquifer's formation; much of the rest trickled in as groundwater flow while the aquifer was still connected to its sources in the snow-covered Rocky Mountains. Now that connection has been severed—save in one small place—and except for the inconsequential rains, inflow has ceased. Water is being pulled out much faster than it is being put back in. Since widespread irrigation began in the 1950s, the Ogallala has sustained a net loss of as much as 120 trillion gallons—11 percent of its original volume. One entire Lake Erie, plus a little. Gone. Most of it gone with full knowledge that it was going. Groundwater overdraft is not an accident here; it is a way of life. But because it means that the water will someday disappear, it is also a way of death.

It would be easy to twist that last sentence into a demand for an immediate end to all Ogallala-sourced irrigation. It would also be wrong. There are a number of very good reasons to continue to pump water out of the Ogallala Aquifer, even though we know it cannot be replaced on anything less than a geologic time scale. The best of these reasons was famously articulated a quarter of a century ago by Steve Reynolds, then New Mexico's state engineer, in an interview with *National Geographic* writer Thomas Y. Canby. "There's nothing intrinsically evil about mining groundwater, as long as everyone understands just what he's doing," Reynolds pointed out. "The alternative is to leave it underground and simply enjoy knowing that it's there."

Groundwater is a mineral, and like most minerals it has practical value. Mining it is a means to realizing that value. We can use mined water to feed ourselves, and clothe ourselves, and keep a substantial portion of the American economy humming—one might say, floating—nicely along. We would be improvident not to do this. But do note the state engineer's caveat: *as long as everyone understands just what he's doing*. Pumping the Ogallala dry will have consequences. It is necessary to understand these consequences, to mitigate them where mitigation is possible, and to figure out how we are going to live with them where it is not. If we can neither mitigate the consequences nor adapt to them, we had better know it now, while it is still possible to turn off the pumps in a controlled manner and keep a little water in the aquifer to temper the problems we will leave our children to deal with as we move on to whatever must inevitably come afterward.

Some of the consequences of groundwater mining are environmental: springs dry up, rivers diminish, the numbers and varieties of plants and animals are reduced. Some are economic: increased pumping costs as wells deepen, increased food costs and decreased land values as crops shrink. And some are human. The human costs may include bankruptcies, foreclosures, and forced migrations. They may include failed businesses and abandoned towns. They are not likely to include thirst—municipal water systems will be among the last users of Ogallala water—but they may well include starvation.

This last will probably happen to people in other countries, not to us. The United States is a wealthy nation, and we can afford to import food as needed. This should not deaden us to the fact that it will be real human lives that are lost. Nor should it blind us to our own peril, should this come to pass. We will be taking bread out of the mouths of other peoples' children. Wars have been fought over much, much less.

Thus the management of water on the High Plains looms as one of the most important American challenges of the twenty-first century. We must find answers to a number of vexing questions. Some are questions of allocation: Once the water is pumped, who receives it? In this, farmers compete with each other, municipalities, industries, recreational users, and wild creatures. Others are questions of quality: Will the water remain safe to drink, to give to livestock, to use on crops? There are debates over the wisdom of water-saving technologies that often lead, paradoxically, to more water use rather than less, and over who should pay for the technologies we choose to employ. Profiteering has become an issue, particularly in Texas, where antiquated water laws allow well owners to suck water from beneath their neighbors' land and sell it to the highest bidder. And the problems posed by interbasin water transport loom larger and larger, as Ogallala users attempt to find replacements for the water they have lost, and as residents of even drier regions attempt to take the remaining water in this underground ocean away.

ALL OF THESE ISSUES can seem grimly unreal as you drive across the green, well-irrigated, apparently endless spaces of the High Plains. You come nose to nose with them at Optima Dam. The phantom reservoir behind it is not the result of poor design, or poor data gathering, or poor construction: The absence of water is directly attributable to groundwater decline. Baseflow—the flow of a river in dry weather—depends on springs, and springs depend on the presence of groundwater. Optima Dam was clearly designed for a baseflow level that is no longer present. Ron Bell, the chief of the Water Management Section for the Corps' Tulsa District, confirmed my suspicions in correspondence a few months after I had visited the place. Here is what Bell had to say:

> The lack of inflow into Optima Lake since its construction is probably due to a combination of factors. The amount and intensity of rainfall events may have declined. Also, farming practices may have had an impact on runoff. I am, however, of the opinion that a decrease in the upper level water table (which may or may not have been affected by the decline in the Ogallala) has probably had the greatest impact.
>
> During the period of record that the dam and lake were being designed, the baseflow at the dam site was usually greater than 20 cubic feet per second (cfs). Rainfall events would then add substantially to the flow in the river. The baseflow since completion of the dam has been zero most of the time. Now when rainfall

does occur, the river channel is bone-dry and most of the runoff is lost to infiltration.

During the period from 1940 to 1966, the average annual flow volume at the dam site was greater than 40,000 acre-feet per year. The average annual flow volume from 1977 to 1993 was only 4,400 acre-feet per year. The flow volume from 1984 to 1993 was only 2,200 acre-feet per year.

It is apparent that something has changed the baseflow and runoff characteristics of the Beaver River.

Note that Bell pointedly refrained from speculating that the drop in the water table that killed Optima Lake was caused by pumping from the Ogallala Aquifer. But in this part of Oklahoma, there is no other reasonable cause.

IF YOU ARE A LOVER of big skies and bigger horizons, travel on the High Plains in summer is about as good as life can get. Clean, black-topped highways thread among green fields; blackbirds declaim from cattails in roadside ditches; hawks preen and glare on fence posts. In the early morning, with the air-conditioning off and the car windows open, you can smell sprinklers and hear meadowlarks; in the late afternoon, thunderheads loom in the long light like galleons. Small, shaded towns appear at discrete intervals, each with its trim frame homes and sandstone business blocks and the dappled green restful-ness of parks where teenagers flirt around community swimming pools and small boys on bicycles eat ice cream beneath giant cotton-woods. It is a perfect Norman Rockwell vision, but it is built on a lie. These fields, these towns, this American-dream lifestyle all depend on the assumption that the water will be available forever. It will not be and it cannot be. Ogallala water is the keystone of an economic and social structure that is poised, as a geologist in Kansas once remarked to me, to "go down like Rome" as soon as the water is gone. We cannot save this lifestyle—not in this form. All we can do is plan for its fail-ure. And we had better be doing exactly that.

What will our water-guzzling culture look like when the water is gone? Can we replace it? Will we find a way to do without? Must we choose between bread and beef, between industry and agriculture, between having enough to eat and enough to drink? If you never miss the water 'til the well runs dry, and there is no future for the well, what is the future of those who depend on the water?

II

THE FLATTEST LAND

O N A MAP, the outline of the Ogallala Aquifer resembles an uprooted mushroom. The cap sprawls across Nebraska, spreading raggedly into Wyoming and edging upward a bit into South Dakota; the knobby stalk extends all the way to the bottom of the Texas Panhandle. Various hyphae reach into New Mexico and Oklahoma and Kansas and Colorado. This description also serves to define the High Plains. The plains and the aquifer are tied intimately together: they were formed at the same time, by the same forces, and they are largely coterminous with one another.

The High Plains share the general geography of the rest of the American Great Plains, but they put their own distinct stamp on it. You know when you have entered them. The land rises, then flattens to a long horizon. The rise may be small, as it is at the thirty-foot step between Brush and Akron in Colorado; or it may be huge, as along the thousand-foot-tall Caprock Escarpment east of Amarillo, Texas. It may be near-vertical, as it is at Caprock, or at Coronado's "palisade," the Mescalero Escarpment of New Mexico; or it may be such a gradual slope, as at Smoky Hill Valley in Kansas, that it is difficult to tell you are climbing. The elevation change is always there. It is at once the source of the name "High Plains" and the key to understanding them.

The broad, flat land that stretches out before you as you top the rise is the surface of a vast geological trash heap, a pile of debris brought down from the ancestral Rocky Mountains by long-vanished streams. It was dumped on top of the western portion of the Great Plains over an age lasting from roughly twenty to three million years ago. Now it is slowly eroding away.

Think of a river coming down a mountain. It is moving fast, over a series of rapids and riffles and waterfalls, and it is carrying pieces of the mountain with it. Some of these pieces ride in the water and are called "suspended load"; others, too big to ride, are rolling along the bottom of the channel—"bed load." When the river is swelled with snowmelt in the spring, you can hear the pieces knocking and grinding against each other, the fine hiss of suspended sands and soils singing a complex counterpoint to the church-organ diapason of the bed load amid the rush and gurgle and thunder of falling water.

Now think of this same river reaching level ground. Its pace slows to a lazy crawl; it no longer has the energy to roll boulders along its bed, or to carry anything much larger than particles of silt. The pieces of mountain it has been carrying fall to the bottom of the channel and stay there. When the channel becomes choked with pieces of mountain, the river moves sideways and cuts a new channel; when the new channel becomes choked, the river moves again. Over and over and over. Channels split, recombine, split again; floods pick up stones and soil and put them down again a mile away, or a hundred miles away. This process is called *anastomosis*, and it is how the High Plains, and the Ogallala Aquifer beneath them, were built. Don't be fooled by talk of underground rivers, of underground lakes, of veins and nodes and dowsing: Aquifers are wet dirt. Water seeps, rather than flows, through them. They do not quickly change. The water that soaks the buried gravels of the Ogallala is largely the water of the vanished rivers that put them there; most of it has been down there for at least three million years. It is sobering, and a little humbling, to realize that the glass of water you just drank was drawn from a stream that vanished about the time our prehuman ancestors first began to walk upright on the ancient savannahs of Africa.

THERE IS A SUPERFICIAL sameness to the High Plains. They have been called the flattest land in the world. South to north, east to west, the land is monotonously horizontal. There are few cities—none, really, north of Texas. There are few watercourses, and those that exist are usually dry. Silos—the skyscrapers of the plains—tower close at hand or loom in the vanishing distance like the masts of ships at the edge of the ocean. The silos are usually white. The land is usually crop-colored: green in spring, yellow in summer, brown with a stubble of dry stalks in fall and winter, like a poor shave several days past. This image is true but incomplete. There is more hidden here than just the Ogallala Aquifer.

Begin with the weather. It is a safe generalization to state that it will be hot in summer, cold in winter, usually windy, and almost always dry. What does that mean? The plains span nearly twelve degrees of latitude: slid sideways to the Atlantic, they would stretch from Savannah, Georgia, all the way to Portland, Maine. They sweep through more than eight degrees of longitude, from the foot of the Rockies nearly through Nebraska. Traveling west across them, you must set your watch back an hour. The climate should not be expected to remain the same over such vast distances, and it does not. Daytime winter temperatures run around 44°F in the south; they drop to half that number in the north. A summer day that tops 100° in Midland, Texas, may not reach 90° in Valentine, Nebraska. In eastern New Mexico, the prevailing winds are almost always from the south or southwest; in northern Kansas they swing around the compass, coming at you from a different direction each season. Even the region's best-known unifying climate factor, its aridity, can vary by surprising amounts. Hobbs, New Mexico, receives a scant twelve inches of rain each year. Norfolk, Nebraska—still on the High Plains, still over the Ogallala Aquifer—gets nearly thirty.

Differences in climate lead to differences in vegetation. Yucca—a towering lily with a woody stem that thrusts upward from a basal mound of sharp, fleshy leaves—is found throughout the region, but it is tall and prolific in the south, stunted and widely spaced in the north. The south has mesquite and indigo bush, two shrubby members of the pea family; in the north, the most common wild members of that family are clovers and low-growing lupines. The rare trees are mostly cottonwood and plum, unless you are in South Dakota or northern Nebraska, in which case they will probably be ponderosa pine. Twinflower, a small honeysuckle that bears two pale-pink blossoms atop an erect four-inch stem, is found only in the north; puccoon, a yellow forget-me-not with large, crumpled, crepe-paper flowers, grows only in the south. The wildflower season starts in the south and rolls north as spring advances. By early April, Scenic Mountain in Big Spring State Park, Texas, is in full, riotous bloom, a natural rock garden of evening primrose, indigo bush, onion, plum, and daisy. South Dakota's Pine Ridge will not reach a similar peak of abundance until mid-May.

I have been speaking of flowers and trees, but the dominant plants of the plains are grasses. These, too, vary with geography. They are tall and lush in the northeast and become gradually shorter and scantier to the west and south. Bluestems, lanky enough to tickle the belly of a tall horse, predominate in eastern Nebraska and Kansas; gramas, barely knee-high to an antelope, take over in the west. Both may be found in-between. Buffalo prefer the gramas (sometimes called "buffalo grasses"), which is why the great herds were found mostly on the western plains.

Grass is meant to be grazed; its leaves grow from the bottom up, putting on new growth at the base rather than at the tip, as other plants do. This simple adaptation has allowed grazers and grass to flourish together. When an animal bites off a flower, the flower dies; when it bites off a grass blade, the remnant of the blade keeps growing. Grasslands are actually healthier with grazers than without. The grass is stimulated to grow; the nutrients in its leaves are recycled through the grazing animal and put back into the ground in a form that makes them available to the grass again. Grazing allows sunlight to reach the earth more frequently, enabling the growth of herbs as well as grasses. Replacing buffalo with cattle has been a step backward for the buffalo-adapted High Plains, but it would be an even greater step backward if there were no large grazing animals at all.

For the traveler in a hurry, the plains are plain. Those in a mood to nose about, though, will find much to savor. There are tall waterfalls draped over sandstone cliffs; there are Kodachrome canyons and cedar swamps and pine-draped ridges. Volcanoes in northeastern New Mexico have capped that part of the plains with lava; erosion in Kansas has eaten all the way through the plains and into the ancient seafloor beneath, leaving pillars and arches of white chalk standing like ghosts of the redrock formations of Arizona and Utah. Once, traveling across flat eastern Colorado beneath a black and threatening sky, my wife and I saw sunlight glinting off bright rock far to the north of our route: A long detour over roads like gravel quarries led us to the Pawnee Buttes, remnants of the High Plains standing by themselves far from the rest. Holes in the ragged sky moved shafts of sunlight across the buttes, highlighting each in turn in the pre-storm darkness, like bright islands rising briefly from a bleak and moody sea.

This land's greatest drama is reserved for its sky: massive, anvil-shaped cumulonimbus, wall clouds hundreds of miles long, rain sweeping across the prairie like the broom of Jove, thunderstorms, tornadoes. An immensity of blue by day, a jewel foundry by night, with what seems like more stars than you can see anywhere else on Earth, save at sea. A rush and flutter of migratory birds. Sunrise and sunset dress themselves in orange scarves and banners; the full moon rises from the flat east like a great wheel, and the long shadows of barns lie like black velvet across the blue cloth of the illuminated night.

The bulk of the plains' surface is taken up by agriculture—cotton in the south, corn and wheat in the north, and alfalfa, sorghum, and pasturage throughout. The landscape these uses create shows its own ordered beauty. Fences line the green geometry of the fields, their top wires holding congresses of birds. Breezes carry the smell of living distance. Culverts lead cattail ponds beneath the highways. These ponds may be fed by springs, or they may be fed by runoff from the fields. The source is the same in either case. That is the Ogallala Aquifer you are looking at. It is this wet presence that allows life to thrive in the middle of North America. We are pumping it out more than three times as fast as Nature can put it back. Over the aquifer as a whole, that loss pencils out to a net deficit of nearly twelve billion gallons a day—enough to supply a typical American family for more than eighty thousand years.

III

PUMPING KANSAS DRY

TWELVE BILLION GALLONS a day is a lot of water. Poured into one-gallon containers and lined up along the equator, a single day's loss from the Ogallala would circle Earth approximately forty-three times. Spread out over the 174,000 square miles of the aquifer, though, it doesn't amount to much: The average daily decline in the water table is about five-thousandths of an inch. That is two inches per year, or twelve feet in the seventy years since serious groundwater use began on the High Plains. The Ogallala has an average saturated thickness—the depth from the water table to the base of the aquifer—of nearly two hundred feet. What in the world are we worried about?

The problem here lies in the word "average." The Ogallala is a highly varied resource. There are places where its saturated thickness is more than one thousand feet; there are other places where it is less than ten. There are places where the water table is dropping more than two feet per year, and other places where it is barely dropping at all. There are a couple of small areas where it is actually rising. Mark Twain, it is said, once remarked that if a man had one foot in a bucket of ice and the other foot in a bucket of boiling water, on the average you would have to say he was comfortable. It is that kind of "average" we are falling victim to with regards to the Ogallala. You cannot trust the general numbers; to get an accurate picture of the aquifer's health, it is necessary to look at specifics. A quick state-by-state survey, south to north:

Texas has 20 percent of the aquifer's land area, but only 12 percent of its water. Drainable water in storage has dropped 25 percent since irrigation development began, and currently stands at about 380 million acre-feet. ("Drainable water in storage" is water currently in the aquifer that can actually be pumped out of wells; it excludes water clinging as a film to rocks and soil particles, water bound into the structure of the soil, and water held in small pores by capillary action.) The water table has declined, on average, forty feet since predevelopment times; it drops another one to two feet each year.

Texas's portion of the High Plains is split in two by the eight-hundred-foot-deep gorge of the Canadian River, which arcs east to west across the panhandle north of Amarillo. Groundwater pumping began south of the river, near the center of the Llano Estacado, as early as 1910; that early development is now part of an L-shaped band of deep depletion that sweeps north up the eastern Llano Estacado from Lubbock to Tulia, then turns west and sprawls all the way across into New Mexico. Throughout most of this nine-county region, the water table has dropped between one hundred and two hundred feet, and saturated thickness has been reduced to less than half its predevelopment amount. There is a second area of deep depletion near Seminole, on the southwestern part of the Llano Estacado. Over roughly half of the Llano Estacado, saturated thickness is now less than fifty feet, and is approaching the condition High Plains farmers call "too thin to pump"—a saturated zone thin enough that high-capacity pumps can no longer operate in it. A third area of deep depletion is centered near Dumas on the North Plains, the name Texans give to the part of their state that lies north of the Canadian River.

New Mexico has 5 percent of the aquifer's land but less than 2 percent of its water. The state contains the western parts of both the North Plains and the Llano Estacado; it also contains the Portales Valley, a broad east–west depression that, several million years ago, held an ancestor of today's Pecos River. Erosion has broken down the ancient valley's walls and refilled much of its bed, and it can be difficult today to tell where the Llano Estacado stops and the valley starts.

Because the High Plains slope slightly to the east, water has been draining out of New Mexico ever since the Pecos changed course and chopped this part of the Ogallala off from its recharge areas in the Rocky Mountains, perhaps as much as 4.5 million years ago. Nearly 30 percent of the state's portion of the aquifer has no significant saturated thickness at all. Irrigated agriculture is largely limited to the eastern Portales Valley and to northern Lea County, around Lovington and Hobbs. The water table has declined between fifty and one hundred feet in both areas; saturated thickness in the Portales Valley has dropped by more than 50 percent. Statewide, 20 percent of New Mexico's Ogallala water is gone, and the water table has declined an average of thirteen feet.

In *Oklahoma*, the High Plains occupy almost all of the panhandle and bulge out into the main part of the state in a half-circle that runs roughly from the Sweetwater River at the Texas border to the Forks of the Cimarron just below Kansas. In the 1930s, this was the heart of the Dust Bowl. Abandoned, collapsing farmsteads from that era interrupt today's fields. The fields are green: Despite water tables that have declined by as much as 150 feet—that is what prevented Optima Lake from filling—most of the panhandle still contains two- to four-hundred feet of saturated thickness. Oklahoma is one of the aquifer's bright spots. More than 90 percent of the 118 million acre-feet of Ogallala water the state started with is still down there.

Kansas is not so lucky. Its figures are similar to those of Texas: 18 percent of the Ogallala's land area, only 10 percent of its water. Nearly sixty million acre-feet of water have been removed from the aquifer in

Kansas since irrigation development began, almost half of that since 1980—a net loss of nineteen quadrillion gallons. The water table beneath the western third of the state has dropped by an average of eighteen feet.

There are two major areas of depletion. The larger area, south of Garden City, extends over most of the six southwest counties of the state and includes water table declines of 100 to 150 feet. The smaller area begins near Scott City and extends northwestward through parts of four counties. The water table has dropped less here—between fifty and one hundred feet—but the drop is more serious, because the water is thinner. More than half the saturated thickness in those four counties is now gone; in many places, the saturated zone has thinned below the thirty-foot minimum necessary to sustain a standard irrigation well.

Colorado's problems are similar to those of New Mexico. The state sits along the upstream fringe of the aquifer, and water has been draining out of it ever since this region of the High Plains was severed from the Rockies, roughly three million years ago. Irrigation has sped the outflow considerably. More than 10 percent of the state's historic portion of the Ogallala is now gone, half of it in the last twenty years. The water table, never healthy, has dropped an average of nine feet; in large parts of Yuma and Kit Carson Counties, along the state's eastern border, the decline is closer to fifty feet. Over the majority of Colorado's portion of the High Plains, the Ogallala's water today is less than fifty feet thick.

Nebraska is a special case. Nebraska has just over one-third of the Ogallala's total land area, but it has nearly two-thirds of the water. It is the only state in which the aquifer is actually gaining: Since 1980, the water table beneath Nebraska has been creeping upward by an average of 1.6 inches per year. Most of this is due to irrigation with water diverted from the Platte River. Water from irrigated fields trickles down through the soil into the Ogallala, providing extra recharge in a broad band extending from the city of North Platte east and south to the town of Holdrege, on the tableland south of the river's valley.

But even Nebraska is not without problems. In Perkins, Chase, and Dundy Counties—stacked up along the eastern side of the notch cut into the Cornhusker State by the northeast corner of Colorado—the water table has dropped more than fifty feet, diminishing the thickness of the saturated part of the aquifer in those counties by an average of 25 percent. Falling water tables have also afflicted the Box Butte Table in the center of the state's panhandle, near Alliance. And although Nebraska boasts the Ogallala's thickest water—as much as twelve hundred feet of saturated thickness, in Grant County—almost all of that water lies under the largely unirrigable Sand Hills. The state's statistics are splendid. The actual situation is somewhat less rosy than the numbers make it appear.

Wyoming's part of the Ogallala is an extension of Nebraska's, and its numbers are similar. In some parts of the state's relatively small section of the aquifer, saturated thickness approaches one thousand feet. This is the highest part of the High Plains, more than a mile above sea level, and the winters are long and cold. Because of this, irrigation has been slow to develop; as recently as 1980, there was no significant agricultural use of the Ogallala's water within the state. That has now changed. In the twenty years between 1980 and 2000, the water table beneath Wyoming's part of the High Plains declined by an average of three and a half feet. No part of the state is yet in difficulty, but the handwriting is on the wall.

South Dakota has just 4,750 square miles of the aquifer—about 3 percent, less than any other state—and a little over 2 percent of its water. Almost none of it has been used. The water is thick: over 100 feet in most places, over 400 feet in many. These statistics are deceiving. South Dakota's portion of the Ogallala is much less healthy than it appears.

The High Plains edges into South Dakota in an arc that runs from Fairfax on the east to Pine Ridge on the west; the top of the arc comes almost to Interstate 90, at Belvidere. This part of the plains is called the Keya Paha Tablelands. It does not give up its water easily, and when it does, the water tends to run high in minerals; high enough, in some cases, to make it unpotable. Human-caused pollution has added to the problem. Other places on the High Plains have talked about importing water to make up for deficiencies in the aquifer; in South Dakota, they are actually doing it, pulling water out of the Missouri River at Fort Pierre and pumping it south. Four hundred feet of saturated thickness is not enough if the water cannot be used once you lift it to the surface.

ONE WAY TO SUMMARIZE the preceding few pages is to state that the Ogallala Aquifer has regional problems, but that it is in good shape overall. Another way is to state that the aquifer has a single problem: water and development can seldom be found in the same place. The first summary is more common; the second is more accurate, and more telling. Almost everywhere development has occurred on the High Plains, the water table has dropped. Often it has dropped dangerously low. The problem we face is not that we will pump the aquifer dry, but that we will pump southwest Kansas dry, and pump the Llano Estacado dry, and pump eastern Colorado dry. The fact that the Nebraska Sand Hills are very likely to escape this fate has little or no practical value. Where development comes in, the water goes out. It is disappointing, but not really surprising, to discover how often this basic association has been overlooked.

Ten miles northwest of Lubbock, Texas, out in the middle of the Llano Estacado, huddles a small town called Shallowater. When the place was founded in 1913, the name was reasonably accurate: Springs fed by Ogallala water still ran in Yellow House Draw, a few feet below the level of Main Street. Today the static level in the municipal wells, the level at which water will stand in the wells when they are not being pumped, is more than one hundred feet below the surface. It cannot help Shallowater—or Lubbock, or Garden City, or Dumas, or any of the High Plains' too-numerous trouble spots—to declare that, on average, the Ogallala Aquifer is comfortable.

IV

DECLINING FORTUNES

I T I S E A S Y to become alarmed at what may happen to the High Plains, and to American food production, when the Ogallala Aquifer runs out of extractable water. It is harder to see what is happening now, before the water is gone. Irrigating crops with groundwater is not an on-off, light-switch event; irrigation will not run full-tilt to a certain point and then suddenly stop. It is more like a gradually closing spigot. As water tables slowly decline, fortunes slowly decline with them. Plenty of water remains, but the pinch is already being felt.

Farmers refer to the cost of pumping water out of the ground as the "lift cost." This cost goes up as the water table goes down. One recent Texas study found that a pump running on natural gas increased its fuel consumption by 26 percent with a one-hundred-foot drop in the water table, going from 7,524 cubic feet of gas per acre-inch of water 150 feet below ground level to 10,224 cubic feet per acre-inch at 250 feet. At three dollars per thousand cubic feet of natural gas, that pencils out to an extra thirteen hundred dollars for each acre-inch of water applied to a 160-acre field—and that doesn't include the costs of labor and maintenance, which usually run about 65 percent of the pumping costs, bringing the total added cost to over twenty-one hundred dollars. Since crop prices depend more on demand than they do on production costs, most of these extra dollars probably will not be recoverable when the crop is sold.

"Lift costs are one thing I don't think we have a good handle on," says Ray Brady, a geologist and water district official who works out of White Deer, Texas. "If you start talking about eight-dollar gas, people start talking about cotton, and dryland, and maybe I'm not going to water this year. How much pumping is reflective of that versus climactic conditions, I don't know, but I've heard more than one ex-farmer say, 'I finally realized I spent more than twenty years working for the gas company.'"

Well yield is also affected by water table decline. Yield, defined as the rate at which a well will supply water to a pump, depends partly on saturated thickness. As the water gets thinner, yield goes down. Choices—what crop to grow, what type of irrigation system to use, when and how long to irrigate—narrow with declining yield. When saturated thickness drops below thirty feet, yield usually is no longer adequate to irrigate at all. Investments in irrigation equipment thus become a crapshoot; the farmer bets that the equipment will pay for itself before declining water levels render it useless. "You can't pick it up and move it, and you can't sell it, other than with the land," one Kansas farmer told me. "So there's some thought that goes into it. You've got to make sure that you're going to stay here a while. And hopefully the water will, too."

Increasing lift costs and declining well yield both squeeze a farm's profit margin. This is an incremental process. A slightly larger bite is taken out of per-acre earnings each year, slowly dragging the farmer under. There are three common methods of dealing with this downward spiral. The first is to seek out more profitable uses for the water; the second is to consolidate farms in order to make an adequate income by pasting many inadequate incomes together; the third is to convert the farm to a non-farm use, such as housing. Thus do declining water levels in the Ogallala feed such disturbing trends as the conversion from low-margin staples to high-margin specialty crops, the increase in corporate farming, and the acceleration of suburban sprawl. Activists decry each of these, but they are often the only choices a farmer has.

The Texas study cited above looked only at irrigation costs. The value of the Ogallala as a factor in the overall economy of the High Plains can also be quantified. One early attempt to do this was documented in a 1984 paper by Harry P. Mapp of Oklahoma State University. Looking only at the Oklahoma portion of the aquifer, Mapp found that irrigated agriculture was generating a net return of $39 million each year. That was in 1977 dollars; the equivalent value in 2004 would be $122 million, or $200,000 per square mile.

More recently, the Docking Institute of Public Affairs at Fort Hays State University in Kansas studied the economic impact of the Ogallala on twelve counties in the southwest corner of their state. Using 1998 dollars, Docking researchers found the "total economic impact" of the Ogallala on their study area to be $188.5 million per year, or roughly $80 for each acre-foot of water pumped out of the ground. Depletion of the aquifer was projected to gradually reduce that value, leading to a net loss of $150,000 per square mile over the next twenty years—a debit, region-wide, of nearly $400 million.

That is the economic story. It does not tell the human story. It does not tell of the 129 vacant housing units in Greensburg, Kansas, which has fewer than 900 housing units altogether and which has issued only two building permits in the past eight years. It does not tell of the boarded-up businesses in Hodgeman County, Kansas (population 2,087 and falling), or the poverty rate in Tulia, Texas (19 percent of the population and rising). It does not tell of the gut-wrenching necessity that drives the sale of a farm that has been in the family for multiple generations, a sale that has been postponed until the bills mount and the creditors bay at the door and there is simply no other way out.

Economic studies do not usually speak of these things, but that is what they are really about.

SOMEWHERE IN THE OKLAHOMA PANHANDLE, on a day of green grass and blue sky and little puffy clouds, I came across an abandoned farmstead. Abandoned farmsteads are common in the panhandle, rising like ruined islands in a cultivated sea, but most of them date from the 1930s. This one was recent. White paint peeled from a hollow-core front door; a television antenna clung crazily to the chimney. A modern refrigerator lay on its side in the weed-grown yard. Sitting in my car in the driveway, I felt like a voyeur to a private tragedy that no stranger should have witnessed. The driveway had been left intact, along with the house, the front yard, and a few outbuildings. Young corn covered the rest, surging against the broken-down fence like a green and storm-tossed sea. The corn was vibrant. There was nothing wrong with the farm; it was the money that had failed.

I could picture the family. Young—there was a rusted tricycle in the yard—and probably idealistic. Definitely hard-working. Somehow they scrape together a down payment, buy the land, build the house, put in a well and a center pivot. The crop is bountiful, and the profits almost cover the bills. They plant again the next year. The water is down six inches in the well, so lift costs go up a little. Commodity prices remain level. The money falls just a bit further behind. They buy the kids' clothes from Goodwill and stretch out the service intervals on the old car. Someone calls from the bank and inquires, very politely, about the late mortgage payment. The well drops another six inches. A neighbor offers to buy them out, keep the house standing, keep the land in production. Maybe the first time the offer is made, they refuse. Another year, another six inches off the well, and selling becomes the only option. That, or foreclosure. The family moves to Guymon and the two farms become one. We decry big agriculture and the loss of family farms, but that is how it usually happens. Six inches at a time.

As long as we use the Ogallala, we will pull it down. As long as we pull it down, changes will occur. Because the pulldown is incremental, the changes are incremental, too. There is no sudden threshhold. But the pain is no less real when the thumbscrews are gradually applied.

V

A STATE OF NATURE

L OWERING THE WATER TABLE has deleterious effects on natural systems as well as on agriculture. The biota changes— not just through conversion of native prairie to crops, but through alteration of the native prairie itself. The reasons are not hard to fathom. Animals and plants are adapted to the moisture conditions under which they live, and if those conditions change, the animals and plants must change as well. Many factors have contributed to the weedlike spread of mesquite on the Texas High Plains—overgrazing and fire suppression among them—but the fact that mesquite roots can reach as far as sixty feet into the earth, while the deepest-rooted grasses only make it twenty feet, certainly has had something to do with it. Living things cannot survive on a hope of water; they must have the real thing, or they will die.

IN 1902, HOMESTEADER WILL TAYLOR planted cottonwoods on his farm six miles north of the little cluster of buildings that would eventually become Portales, New Mexico. The trees thrived and grew; Taylor planted a few more. Taylor's Oasis became a popular picnic spot. Familes from Portales and nearby Clovis drove out to the Taylor

place on Sunday afternoons to eat lunch in the shade and play on the
sand dunes that had built up as a result of poor cultivation practices
by some of Will Taylor's neighbors. At that time, the water in the Ogallala Aquifer was only a few feet down.

In 1962, sixty years after the first cottonwoods went into the
ground, Taylor's Oasis was purchased by the State of New Mexico and
transformed into Oasis State Park. The park's primary purpose was
recreation, but the state also wanted to protect the homestead, which
had become unique. Land-use patterns in the Portales Valley were
changing. Small holdings were being combined; single fields stretched
over what had once been whole farms. Irrigation was increasing, and
by 1978 it had become the norm: More than half the farmland in the
valley was now watered from the Ogallala Aquifer. The irrigated portion included Oasis State Park. The water table had dropped well
below the cottonwoods' roots, and the park had been forced to put in
a well to keep the principal reasons for its existence alive.

Oasis State Park still felt like an oasis when I visited it in spring
2003. Nature trails led among the dunes; cottonwoods towered over
campsites and picnic tables. A three-acre lake, added in 1972, lazed
with fishermen and waterfowl. The land glowed golden in the late
afternoon light, like the happily-ever-after picture in a book of fairy
tales. But happily ever after was not on park manager Jim Whary's
agenda—not, at least, without some serious intervention.

"When these cottonwoods were planted, the water was maybe
thirty-six inches down," Whary told me in the park's small visitors
center. "Now in one of our wells the static level is at 125 feet and in the
other, at 85 feet. That says a lot right there."

I asked how the cottonwoods were doing. "Terrible," he responded,
immediately. Part of that is age: Cottonwoods have an average life
span of around eighty years, and a number of the park's trees are
approaching a century. But even the younger ones are suffering from

lack of water, which the park's irrigation system has been only partly able to alleviate. Drip lines to some of the trees have helped, but the help may be too little and too late.

Water availability—or the lack of it—has also impacted the park's lake, which may be the most pampered body of water in the United States. The lake bottom has been lined with bentonite clay to prevent leakage into the ground; as a result, boating, wading, and swimming had to be banned, because they might stir up the bentonite and destroy the seal. The water is edged with square sandstone blocks set in concrete, giving the little lake the appearance of a large swimming pool or decorative garden pond. "The fishermen hate it," Whary smiled sadly. The park staff doesn't much like it either, but they feel they have no choice. The lake is filled with Ogallala water, pumped from the deeper of the park's two wells. Oasis State Park is allotted fifty acre-feet per year, out of which they must not only maintain the lake but keep the cottonwoods alive and provide for the campground, the visitors center, and the fourteen species of mammals, more than eighty species of birds, and at least fifty species of native plants that call the park home. Bentonite and sandstone edging are awkward but necessary.

Lake Meade in southwest Kansas, like the lake in Oasis State Park, is filled with water pumped from the Ogallala Aquifer. There most of the similarities stop. One difference is size; Lake Meade covers eighty acres, while Jim Whary's carefully tended baby in New Mexico spreads out over barely three. A second difference is management style; there is no bentonite bottom beneath Lake Meade, and no garden-pond edge. The waterline is scalloped and natural. One corner of the lake is exuberent with cattails. There is a swimming beach—empty on the April day I visited—and a boat ramp. But the biggest difference probably lies in the lakes' separate histories. In Oasis State

Park, the water has always been pumped. When Lake Meade was built, the water was still arriving on its own.

The basin now filled by the lake was once Stumpy Arroyo, part of a large Meade County cattle operation called the Turkey Track Ranch. It was purchased by the State of Kansas in 1927 specifically to build a lake. There was then no place for waterborne recreation in southwest Kansas, and Stumpy Arroyo, which had a perennial stream fed by numerous Ogallala springs, seemed a perfect place to offer it. The dam went up in the early 1930s, and the lake filled rapidly to the fifteen-foot spillway depth. A hatchery was constructed nearby, fed by an Ogallala well—perhaps the only fish-raising facility in the world where the fish swim in three-million-year-old fossil water—and for twenty years, all was well. Then irrigation came to Meade County, and Lake Meade started to go away.

By 1983 the lake had dropped twelve feet, leaving a three-foot-deep puddle sloshing forlornly about in the bottom of an otherwise dry basin. Park managers arranged for a temporary pipeline from the hatchery's well; to avoid overstressing the well, it was stipulated that only enough water would be pumped to raise the lake to three feet below the level of the spillway. The aim was to stabilize the situation until something better could be found.

Today the "temporary" pipeline is still there, the lake is still three feet lower than the spillway, and "something better" is not even on the horizon. Things will probably remain this way until the well gives out or the lake fills in, whichever comes first. "The reality of this lake's existence," states the 2003 *Kansas Water Plan*, is that it is "tied directly to the Ogallala and will likely require pumping supplementation through the life of the lake, since the Ogallala water table has dropped and no longer provides the needed baseflow." Like thousands of other places and creatures and plants throughout large parts of eight states, Lake Meade's fate is inextricably intertwined with the fate of the Ogallala Aquifer. On the High Plains, nature is no longer natural: All of it, every bit, depends on pumping levels.

A FEW MILES NORTH of Scott City, Kansas, along US 83, the flat infinity of the High Plains gives way suddenly to the two-hundred-foot-deep canyon of Ladder Creek, carved out of Ogallala-formation gravels held together by cementlike accretions of calcium carbonate. A side road branches west and plunges into the canyon. A glint of open water appears. This is Lake Scott, formed in 1930 by the damming of Ladder Creek, shortly after the canyon became a state park. The lake is fed by the creek; the creek is fed by Ogallala springs. One of these, Big Springs, is the home of a creature found nowhere else on earth.

The Scott riffle beetle, *Optioservus phaeus*, is a stubby, nondescript brown insect approximately an eighth of an inch long. It spends its days licking rocks, cleaning off the periphyton, the thin layer of organic ooze that clings to surfaces beneath the water. You might find one useful in your aquarium, but you wouldn't be able to obtain it legally. The beetle was classed as "threatened" under the Endangered Species Act in 1978, the same year that University of Michigan biologist David S. White described it as an independent species. Twenty years later, after an exhaustive but fruitless search for others of its kind in other High Plains springs—including every spring in Lake Scott State Park and adjoining areas—the little beetle's status was upgraded to "endangered." Because of its severely limited habitat and range, it is likely to retain that classification as long as either the animal or the classification exists.

Protection of endangered species is often controversial. This one would not appear to be, at least at first glance. The Scott riffle beetle's entire known habitat is within a state park, which has been protected ground since before 1930; no change of ownership, or even of management goals, is necessary. There are no nearby sources of pollution. The beetle's population is limited in range but large in number, and is therefore resilient. The species has a lengthy reproductive life—individuals have been kept in aquaria for as long as nine years—so even if there were a one-time catastrophic loss of all but a few adults, the population should be able to rebound. The only way to kill off the Scott riffle beetle would seem to be to dry up Big Springs. But that, it turns out, is precisely what biologists are worried about.

The water that feeds Big Springs comes from Scott County's portion of the Ogallala Aquifer. This is the most endangered part of the aquifer in Kansas. The last fifty years have seen the saturated thickness in parts of the Western Kansas Groundwater Management District, which includes Scott County, decline by more than 50 percent; in some places, the declines have been greater than 90 percent, leaving as little as three feet of water still in the ground. More than twenty-eight hundred high-capacity wells—almost two per square mile—continue to draw the water out. The water table is currently falling an average of a foot each year; the district expects to be out of pumpable water by 2025.

What will happen to the Scott riffle beetle at that time is unclear. "Out of pumpable water" does not mean out of *all* water, just out of water available to high-capacity wells; the common criterion is thirty feet or less of saturated thickness. Big Springs flows from the very bottom of the aquifer, out of the contact zone between the Ogallala Formation and the underlying Niobrara shale; when the pumping stops, the saturated thickness above the spring should still be at least thirty feet. Natural recharge will increase the thickness, or at least hold it steady, after that. In all likelihood, the water will continue to flow.

How much of it flows may be another matter entirely. Spring flow, like well yield, depends on saturated thickness, and as saturated thickness in Scott County has diminished, Big Springs's flow has diminished with it. The decrease is still relatively small: from 400 gallons per minute in 1974 to 350 gallons per minute today, a 13 percent drop in thirty years. We cannot predict how much further the flow will atrophy in the next thirty years, but we can say with certainty that it will not increase. Big Springs is on its way to becoming Little Springs. Whether it will remain big enough to keep the Scott riffle beetle alive is unknown, but the signs are not particularly hopeful. The beetle has been there for many, many centuries. There is no guarantee that it will be allowed to survive this one.

LEGITIMATE QUESTIONS CAN be raised about whether preserving picnic groves and fishing lakes and beetle species are proper uses for the limited resources of the Ogallala. In the northern part of the Texas Panhandle, people are raising these very questions. Near Amarillo, a well and pipeline similar to those at Lake Meade and Oasis State Park have been proposed to reverse falling water levels in a small reservoir called Lake McClellan—a solution that troubles Ray Brady, the geologist and water district official from White Deer, within whose district the lake, the well, and the proposed pipeline all lie.

"The district's position on that is that it will constitute a waste of water," Brady told me as his pickup rattled across a McClellan Creek bridge a few miles upstream of the reservoir. "We did a few back-of-the-envelope calculations. What's the local evaporation rate? What's the surface area of the reservoir? How many acre-feet go out versus how many you're putting in? You're talking about roughly a 30 to 35 percent loss if you take groundwater and discharge it to a surface reservoir. I don't think that qualifies as efficient use."

Brady is undoubtedly right about the efficiency of fishing lakes. The question remains whether efficiency is always the right criterion to judge by, and I am not at all sure that it is. When everything is dependent upon a single resource, everything needs to get a little bit of it. Efficiency—or, to turn that around, waste—may be a good measuring stick to help determine who or what, among users of the same type, should get the water, but it fails miserably when the choice is among different types of users. It is especially misleading when it tries to deal with the needs of nature. The highest waterfall in Nebraska, an exquisite seventy-eight-foot-tall curtain of flowing traceries down a red sandstone wall in the canyon of the Niobrara River, falls from an Ogallala spring. It is mindlessly wasteful: In its short run from spring to river, no one does anything with the water except look at it. There are no doubt people who would happily plug it to preserve agriculture on the canyon's rim, but most of us would not call them sane. Where a state of nature no longer exists, preserving a reasonable facsimile of it can sometimes be the best possible use of a scarce resource.

Big Springs is actually healthier than most other springs on the High Plains. Some have dried up entirely; others have been reduced to intermittent flows and seeps. In a paper published in 2002, Robert S. Sawin of the Kansas Geological Survey reported on visits to fifty-seven western Kansas springs that existed in historic records; of these, only five were flowing at unreduced rates, and eight—including some with long histories of human use—were nearly or completely dry. Things are even worse in Texas, as can be seen by leafing through *The Springs of Texas, Vol. I*, geologist Gunnar Brune's detailed 1981 examination of most of the springs, seeps, and damp mud banks of the Lone Star State (it would have been all of them, but Brune died before the smaller second volume could be completed). "The story of Texas's springs is largely a story of the past," Brune wrote, and the accounts in this massive self-published book bear that statement out. One reads of favorite pioneer picnic grounds, of fish-filled lakes and creeks, of powerful rapids and dancing waterfalls—all gone. Spring Lake, in Lamb County, was deep and clear and "very popular for swimming and fishing" in the 1920s; when Brune visited it in 1978, it had been reduced to "a small pool of runoff water" from a stock-tank windmill. Silver Falls, in Crosby County, once consisted of "a large volume of crystal-clear spring water passing over . . . sandstone ledges." Brune found it dry. Hale County had no springs left at all. At Greene Springs in Scurry County, seeping from a detached section of the Ogallala Formation a few miles off the eastern edge of the Llano Estacado, the geologist noted water slides carved either by or for pre-Columbian children, complete with carefully cut steps and smooth channels just the width of a child's bottom. They trailed forlornly down a dry sandstone face over which a few drops of water trickled, many feet away from the play area. Brune had no doubt about the cause of these and similar losses: The overwhelming majority of them, he wrote, were a direct result of irrigation pumping. Graphs included for some of the springs in the book bear this out, showing steady flows until the first wells come in and declining flows thereafter, with the steepest declines coming during the times of most rapid nearby development.

Rivers, too, are failing. The Kansas portion of the Arkansas River, deprived of baseflow by dropping water tables and bled off by irrigators in Colorado, has long since ceased to exist above Great Bend. The Republican River in Kansas, Colorado, and Nebraska is much reduced, and the three states are fighting over the pittance of water that remains. The average annual flow of the Platte River near its confluence with the Missouri is only 16 percent of its volume during the first years of record, in the middle of the nineteenth century. These rivers' anemic conditions have usually been attributed to diversion for irrigation, but water table decline is at least as significant. Historically, most plains rivers were gaining streams—streams that obtained more groundwater from springs in their beds than they lost through exfiltration (the seepage of water out of the river into the earth). This status has been altered by groundwater pumping in two ways. Less saturated thickness means less spring flow for gaining streams to gain; a lowered water table means that more of the rivers' beds now lie above the saturated zone, converting long stretches from gaining to losing. An atrophied river is the result. Water managers worry about efficient use of water. Inefficient uses may be equally important.

Ray Brady would probably agree with that statement; like most water managers I met on the High Plains, he is at least as concerned about using water well as he is about using it efficiently. "What's the value of water?" he asked rhetorically in Texas, as the road to Lake McClellan receded in his pickup's rear-view mirror. "I'm not sure what the value of water is right now. I think maybe we worry more about the price of water than about its value."

I told him about the care they were taking not to waste water in Oasis State Park's lake. Brady grinned wryly.

"Well," he allowed, "there must not be many other places to fish around."

VI

DON'T DRINK THE WATER

THERE WAS A TIME when groundwater was considered inviolate. Locked safely in its dark fortress beneath the earth, thoroughly filtered by the damp soil, it was assumed to be immune to the quality problems that plague surface water. We know better now. Groundwater is harder to pollute than surface water, but the job can be done, and when it is, the mess that results is much more difficult to clean up.

The Ogallala Aquifer, as a whole, is not polluted. What you are about to read is a group of little horror stories, not one big one. Overall, the waters of the Ogallala are in excellent shape: a little mineralized, perhaps, from three million years of sitting around in the earth, but fresh, clean, and perfectly safe to drink. Most municipal water systems that use Ogallala water do not even bother to treat it, beyond adding a little chlorine at the wellhead to protect against possible bacterial growth in their distribution systems. Some do not do even that. "The only time we inject chlorine into our water is on the rare occasion that a water sample tests positive for coliform bacteria," water supervisor Mike Hulquist told me emphatically in Alliance, Nebraska. "We do absolutely no other treatment of any kind."

But if the general picture is good, the specifics are not always quite so rosy. The decline in the Ogallala's quantity has been accompanied, in many places, by a decline in its quality. Agricultural chemicals have been found in roughly one quarter of the wells that have been tested for them in Kansas, Texas, Oklahoma, and Colorado. Seventeen small portions of the aquifer, scattered over four states, are contaminated badly enough to qualify for Superfund status. And where suburban sprawl has reached the plains, it has proved to have impacts below ground as well as above. "Don't drink the water" is no longer advice that is restricted to surface supplies; it is increasingly applicable to water pumped from the once "inviolate" earth.

OF THE EXAMPLES just cited, agricultural contamination is the most troubling. Not because the concentrations are high (they are not), but because the implications are enormous. Agriculture is not just the backbone of the High Plains economy, it *is* the High Plains economy. The 174,000 square miles of the plains—just 5 percent of the land area of the United States—produce 30 percent of the nation's irrigated agriculture and 40 percent of its beef. Ninety-five percent of the water pumped from the Ogallala Aquifer is used for irrigation. More than 99 percent of the land in some High Plains counties is classified as farmland. The plains are not a bioregion; they are a food production facility a quarter of a continent wide. "Don't be fooled by all the fresh air and sunshine," remarks *National Geographic* writer Erla Zwingle: "This isn't 'landscape' any more than an office or a factory is."

Water use and chemical use are closely paired in irrigated agriculture. Fertilizers and pesticides are commonly applied with the irrigation water, a process called "chemigation." This approach has several advantages over other methods of getting the chemicals to the crop. Far less of the material enters the air with chemigation than it does with aerial spraying, broadcast spraying from tractor-drawn equipment, or even hand application. There is minimal human contact. And because the chemicals are carried in the water, they go where the water goes, which is generally where the farmer wants them: into the soil for fertilizers or nematode control, onto the leaves for rusts or aphids or the destruction of broadleaf weeds. Better targeting means that smaller amounts can be used, benefiting both the environment and the farmer's pocketbook. Harm to nontarget organisms can be reduced or avoided altogether.

But this advantage is also chemigation's major disadvantage. The chemicals go where the water goes. If the right amount of water is applied, that is only into the soil and onto the plants, but if there is more water than the soil and the plants can absorb, it has to go someplace else. There are two possibilities. One is into a body of surface water—a lake, a pond, or a stream. The other is downward, toward the water table.

Farmers try to avoid applying excess water. Water costs money to pump and to distribute, and any that is wasted translates directly into wasted dollars. Farmers, like all good businesspeople, try to minimize those. But there are fourteen million acres under irrigation on the High Plains. Even if there was only a one-in-a-million chance of overwatering any given acre on any given day, overwatering would still happen somewhere on the plains fourteen times a day.

And the actual chance of overwatering is considerably higher than one in a million. I am aware of no figures that track it, but an educated guess would put it closer to one in ten. That is because determining the water capacity of the agricultural system—the plants and the soil—is an inexact science. A plant's needs depend both on its transpiration rate (the amount of water exhaled from the leaf pores) and its metabolic rate (the speed at which food is converted into energy and new plant matter). Both transpiration and metabolic rates vary with temperature, humidity, season, and time of day. The soil's ability to hold water, called "field capacity," depends on the materials it is made of, the way those materials are put together (loose, clumped, or compacted), the current evaporation rate, the soil moisture tension (the amount of water already in the soil), and a host of other factors, not all of which can be accurately measured. The farmer measures what he or she can, guesses at the rest, and hopes for the best. But the crop suffers if the guess is low, so the tendancy is to give hope a nudge by guessing just a little bit high.

The problem of agricultural chemicals in groundwater is still very minor. The most thorough study I am aware of—a 1999 survey of the central High Plains by Mark Becker, Breton Bruce, Larry Pope, and William Andrews of the U.S. Geological Survey—found agricultural chemicals in slightly less than one quarter of all wells tested, always in only trace amounts. "All of the detected pesticides that have [EPA-assigned] Maximum Contaminant Levels were measured at concentrations substantially less than those levels," the USGS team wrote. You shouldn't stop drinking the water from the municipal wells of Garden City, Kansas, or Guymon, Oklahoma, or Holdrege, Nebraska, just because these cities are surrounded by farmers who chemigate with pesticides.

But there remains something troubling about broad-scale agricultural chemical use on the High Plains. Groundwater recharge is a geological process; it is rapid as geology goes, but agonizingly slow on a human scale. Chemigation is new: a nanosecond long, geologically speaking. The concentrations of pesticides and fertilizers in the Ogallala's water are low today. We cannot predict that they will remain low fifty years from now. What the USGS team measured may be simply the first wave of an incoming ocean.

Or it may not be. The future is notoriously hard to predict. The Ogallala Aquifer is huge; its size alone may be enough to buffer it against serious problems. The slow pace of groundwater movement may give pesticides, which are often unstable, time to evolve into less hazardous materials before they spread very far. Clay is both a barrier to groundwater flow and an adsorbent for chemicals, and there are beds of clay scattered here and there throughout the aquifer. It is not unrealistic to hope that contamination by agricultural chemicals will never be more than a minor nuisance.

But public policy cannot be built on hope. Policy-making requires sound science, projectible trends, and an ability to balance the individual good of each segment of society against the general good of all of us. And when you look at agricultural contamination of the Ogallala Aquifer through that set of lenses, two disturbing facts emerge. First, the contamination is widespread: It is found in all regions where it has been looked for, though not in all wells and nowhere in more than minimal amounts. And second, the contamination is shallow: Agricultural chemicals have been found almost exclusively in the uppermost part of the aquifer.

That second point seems particularly troubling. In the 1999 USGS study, nearly 80 percent of the contaminated wells were less than two hundred feet deep. Of wells more than three hundred feet deep, only three showed chemical contamination. This could indicate that the aquifer is protecting itself against deep contamination. More likely, it means that the contaminants simply haven't got down there yet. The chemical buildup may have just begun. A slow creep downward through the soil to the water table; a slower diffusion through the saturated zone toward the base of the aquifer. And on top of the system, out in the air, blissfully unaware of what is happening beneath our feet, we continue to apply more chemicals—in forms deliberately designed to be carried by water. We are a clever race, but we are not always very wise.

AGRICULTURAL CONTAMINATION of the Ogallala Aquifer is characterized by very low levels, but very broad scope. The seventeen Superfund sites found over the aquifer have characteristics that are just the opposite: high contaminant levels, but very limited scope. They are of serious concern where they are found, but they are not found widely. And while they are expensive and time-consuming to clean up, cleanup is almost always an option.

Superfund—officially, the Hazardous Substance Response Trust Fund, but nobody ever calls it that—is a revolving account set up by Congress in 1980 to deal with the presence of hazardous substances in places where they do not belong. Most of the money in the fund comes from taxes on the manufacture and distribution of the hazardous substances themselves. Because the fund was set up in response to the discovery of some particularly egregious illegal waste dumps (notably, Love Canal near Niagara Falls, New York, and the Valley of the Drums near Louisville, Kentucky), people usually associate it with the spectacular: acres of decaying steel drums, miles of abandoned factories, rivers that burn, lakes whose water kills ducks on contact. Such sites do exist, but they are relatively rare. Most places that make the National Priorities List, the collection of hazardous-materials sites that are eligible for Superfund cleanup, are pretty low key. There must be a public health hazard or an environmental threat involved, or listing cannot be triggered. Beyond that threshhold criterion, though, the real test is not how big the threat is, but how likely it is to be cleaned up without government assistance.

The Ace Services site in Colby, Kansas, is typical of the Superfund sites found over the Ogallala. Ace Services was a small-scale machine shop and metal-plating facility established in 1969 on the eastern edge of downtown Colby that primarily worked on farm machinery. The shop was built on a weedy, two-acre lot beside a branch line of the Union Pacific Railroad, near a creekbed that almost never had a creek in it. It is unlikely that the company's proprietors were trying to get away with environmental rape when they rid themselves of their plating wastes by dumping them on the ground behind the building; they probably thought there was no environment left on that property to rape. A reasonable assumption, if chemicals would only stay where put.

In 1971, just two years after the plant opened, someone in Colby complained to the Kansas Department of Health and Environment

(KDHE), and Ace was ordered to stop dumping its wastes on the ground. The company complied, constructing a wastewater lagoon and a pair of large concrete vats, but the work was done hastily and incompletely. The lagoon was unlined, the vats were unsealed, and everything leaked. Plating wastes continued to sink into the soil and percolate downward. Sometime in the mid-1970s they reached the Ogallala Aquifer.

Pollution in an aquifer spreads slowly in the direction of groundwater flow, forming a fan-shaped region of elevated contaminant levels known as a "plume." Groundwater flow beneath Colby, as in most parts of the Ogallala, is generally toward the east; the Ace Services plume spread eastward. By 1980 it had reached Colby Municipal Well Number 8, a thousand feet from the plant, where it showed up as levels of chromium as high as eighteen hundred parts per billion— eighteen times safe drinking-water standards. The well was closed, and KDHE once again went knocking on Ace's door.

The machine shop installed a wastewater treatment system and drilled a recovery well into the plume to intercept the flow of polluted groundwater and pump it out. The technology was expensive, and it didn't work; its main effect appears to have been to bankrupt the company. In 1989, Ace Services went out of business. Tests for chromium in the soil at the abandoned site found accumulations of as much as 19,100 parts per million. Private wells in the plume, which was now a mile long and a quarter of a mile wide, showed chromium levels up to forty-one times safe drinking-water standards. The wells were capped and their owners' faucets were connected to the city water system. The shop site went onto the National Priorities List.

As I write this in fall 2004, the cleanup is well underway. The buildings on the site have been demolished and the rubble has been hauled off and isolated in a lined landfill, along with more than three thousand cubic yards of contaminated soil. Pumps are attacking the plume, bringing it up through extraction wells and running it through a treatment plant with a resin filter designed to remove heavy metals; the cleansed water, now safe to drink, is dumped into Colby's city water system. "The resin treatment plant has worked great," EPA site manager Robert Stewart assured me in a September 2004 e-mail. "It delivers water that easily meets the cleanup goal." The plume is shrinking; Stewart expects it to disappear entirely by 2017. At that point, the EPA intends to pick up its resin plant and go home.

All of which sounds like a success story. It *is* a success story. Contamination of the aquifer has been caught, arrested, and cleaned up. Colby has safe drinking water again. The laws, and the technology, work. Perfect—except for a couple of nagging worries.

The first nag concerns money. The Colby cleanup has cost around $8 million to date; when the work is complete, the total costs are likely to be at least twice that. This works out to around one hundred thousand dollars per acre of cleaned-up plume. There are 111 million acres above the Ogallala Aquifer. If even one-half of 1 percent of that total acreage needs cleaning, and if costs remain steady (they never do), we are talking about at least $555 billion.

But we won't have to spend that much, right? The contaminated areas are few and small; only seventeen Superfund sites in the entire 111 million acres. If Colby is typical, we are only talking about 0.002 percent of the aquifer's surface area. Perhaps half of a percent is exaggerating?

Perhaps. But that is where the second nag comes in. It enters obliquely, stage left, in something that Robert Stewart reported from Colby. During the site's semiannual checkup in June 2004, Stewart told me happily, "the plume was being captured adequately by the extraction wells, even though at that time several wells had been turned off awaiting construction by KDHE of a granular activated carbon treatment plant to treat volatile organics from a LUST site nearby that had been drawn to our interior wells."

He didn't translate that second acronym; I had to look it up. It stands for Leaking Underground Storage Tank, and those, it turns out, are all over—at abandoned gasoline stations, at active gasoline stations, at homes and businesses that heat with oil, on tank farms, on real farms—you name it, it probably has or has had a LUST site that could be drawn to its interior wells. And LUST sites don't usually show up on the National Priorities List.

There are numerous ways, large and small, by which we are polluting the Ogallala Aquifer. We put cattle feedlots over it; we spray pesticides and fertilizers onto it; we drill oil wells through it. Near Amarillo, Texas, we build nuclear weapons on top of it. We don't even know what we are doing to it in the little unincorporated village of Wright, Kansas; all we know is that volatile organic chemicals have begun to show up in the residents' wells. Life creates by-products, and living as we do creates particularly nasty ones. If they escape from the places we put them, they can do a great deal of damage before we manage to pen them up again.

And though it is easy to raise a panic over them, the particularly nasty ones may not be the worst. At least they get our attention. Perhaps it is more accurate to say that they divert our attention. Pollution can come to groundwater in many forms. We don't have to make volatile organic chemicals, or fissile materials, or chrome residues, or anything else that is frightening and obscure in order to endanger the Ogallala Aquifer; all we have to make is something that all of us produce at a rate of roughly half a pound every day. It may not be as glamorous as Pentachlor or plutonium or 2,4,5-T, but human body waste can waste us every bit as badly.

THE ASSUMPTION THAT groundwater is inviolate was based largely on our experience with microorganisms, which can be filtered out by the earth. That is one basis for the requirement, in modern wilderness etiquette, to defecate at least two hundred feet from a lake or a stream—a rule that seemed to have plenty of leeway built into it when it was formulated thirty years ago. It was about twenty-five times the distance sanitation scientists thought was perfectly safe. The claim made by the authors of one popular textbook *Water Supply and Pollution Control*, published in 1977, was typical of the thinking of that time:

> The number of harmful enteric [intestinal] organisms is gener-
> ally reduced to tolerable levels by the percolation of water
> through 6 or 7 ft of fine-grained soil.

And perhaps it is, if the soil is fresh enough and the word "tolerable" is defined broadly enough. What has been realized since the 1970s, though, is that soil—like any system—can be overwhelmed. Flood enough bacteria onto six or seven feet of soil, or even sixty or seventy feet, and some of them are going to break through. Good aquifers are not themselves good filters, because the pore spaces are too large. Once bacteria get down there, they—like chemigation chemicals—are going to go wherever the water goes.

The problem is most acute in areas plagued by suburban sprawl. The houses spread out and the septic tanks dig in, drainfield upon drainfield upon drainfield for mile after mile after mile along the popular commuting routes into the cities. Ray Brady, the Texas geologist and water district official, pointed out some of these string suburbs to me as we drove near Amarillo. "People want to move out into the country," he complained, "so they'll buy these little nine- or ten-acre tracts—house and a well, house and a well, house and a well—you see a lot of that southwest of the city. And water consumption may not be that big of a deal. But I'm more concerned about quality, because every one of those has got a septic system."

Septic systems work very well—when they are properly maintained. When they are not, they are no more sanitary than an outhouse, quite possibly less so. My wife and I once lived in a house in

Oregon with a failed septic system. After we had been in the house for a while, we discovered that everything that disappeared when we flushed the toilet was reappearing, like magic, under a rosebush in the backyard. As a friend pointed out at the time, the only detectable difference between sitting in the bathroom and sitting in the rosebush was that the bathroom was more comfortable.

"It's kind of a good news, bad news type of thing," Brady points out. "The good news is that you're always getting some aquifer recharge. The bad news is that it's your sewer that's the recharge. Of course, the real estate fellow will not tell them that." He waves a hand toward a row of houses down a side road. "A lot of that used to be cropland. Places where I measured wells in 1998 are no longer planted in milo, they are fully planted in houses. So water use patterns are changing, and I suspect, though I don't know, that pumping quantity will go down. But I wonder about the quality side. I don't know what that's going to be like around here in ten years, but it's not going to be cropland. Amarillo's going to grow, Lubbock's going to grow—not in the cities, but in bedroom communities thirty miles out."

The problem is not the bedrooms, of course. It is the bathrooms that go along with them.

POLLUTION OF THE OGALLALA AQUIFER by human waste is still largely an issue for the future in Texas. It is an issue right now in South Dakota, where a combination of natural mineralization and human contamination has closed so many wells on the Pine Ridge and Rosebud Indian Reservations that the Lakota tribe has begun importing water from the Missouri River at Fort Pierre, as much as sixteen hundred feet lower and two hundred miles away from the places where the water is intended to be used.

"If we stay with the aquifer, we're going to have to go deeper," explains Delano Featherman. "A good deal of the water is contaminated. We treat it as we pull it out, but with all the recent chemicals,

we have to keep monitoring constantly." Featherman is the land acquisition officer for the Mni Wiconi project on the Pine Ridge Reservation, which is tucked into the southern curve of Badlands National
Park just north of the Nebraska border. "Mni wiconi" means "water is
life" in the language of the Oglala Lakota, and that is exactly what the
project will mean here at the northern end of the Ogallala Aquifer,
among the people who have lent it their name. Poverty has teamed with
off-reservation agriculture and unusual soil chemistry to make much of
the reservation's groundwater unsafe for human consumption. The
Oglala can no longer use the Ogallala, and that says things about our
priorities as a society that many of us would prefer not to hear.

"When they spray the fields around us in the spring and the summer, we see these big ol' planes flying around," Featherman continues.
"They're looking at an infestation of grasshoppers, and there's rodents
that utilize the fields out there for their survival, so there's a lot of
spraying going on. It's seeping into the ground, seeping into whatever
shallow strips of aquifer we have down there, and it's contaminating
them." There is also natural pollution—if the term is not an oxymoron—from the soils of the Badlands, which are heavily laden with
volcanic ash. There is runoff from roads and freeways. And if soil
chemistry and agricultural chemistry and modern transportation are
not enough to trouble you, there is also the sewage.

The sewage problems on the Lakota reservations have been compounded by the tribe's success in dealing with its water quality problems—among these, the sewage problem itself. The Lakota have a
dispersed lifestyle; many of them still live rurally, on small tracts
tucked among the trees and meadows and stone outcrops of the Pine
Ridge. When water quality in the part of the Ogallala Aquifer beneath
the reservation began going downhill a few years ago, the private wells
that were relied on for water in the rural areas were forced to close,
and people moved into town. Now that Mni Wiconi is a reality, they
all want to come back.

"You see trailer houses going in all over, and we have to hook them up," Featherman complains. "In the areas where we've already completed our project, there's people moving back, moving back, constantly, and it's hard to keep up with them. And there's the housing projects—there's three different projects going in right now. They're out there setting up their houses, and they all want us to hook them up next."

At nearby Oglala Lakota College, range ecologist Trudy Ecoffey confirms Featherman's analysis. She worries about septic systems. "Indian Health actually puts a lot of those in," she points out, referring to the federally operated Indian Health Service. "They're supposed to do an environmental assessment, but I don't think they're giving much thought to it. And the population around here is growing. Fifty percent of the population on the Pine Ridge Reservation is under the age of eighteen. People want to stay here on the reservation, and they want houses. So they're putting in septic systems, and wells, and not giving much thought to the underground waters."

Featherman explains the operation of the Mni Wiconi system. A treatment plant at Fort Pierre, on the west bank of the Missouri River near the state capital, draws water from the river. Treated water is dumped into a twenty-four-inch pipeline and pumped south. "We've got it as far as Belvidere right now," Featherman notes. "We're supposed to have a big master control in Wanblee—that's all going to be computerized. From there, the lines will run straight west into the town of Pine Ridge." Branch lines will split off to serve the neighboring Rosebud Reservation. At least one branch will continue south beyond Indian lands, all the way into Nebraska. Despite its name, the Mni Wiconi pipeline wasn't originally designed to serve the Lakota at all, although it would have crossed Lakota land. The tribe's current role in the project is the result of some adept political maneuvering on the part of Featherman and others; the original nontribal beneficiaries are still included, and must still be served. One of these is a proposed hog farm that, if built, will immediately become the world's third largest. Hogs create about the same amount of sewage, pound for pound, as humans do. Thus do solutions to problems compound the very problems they were designed to solve.

The Pine Ridge Reservation is the first broad-scale area where water quality problems in the Ogallala have forced changes in the way the aquifer is utilized. It is not likely to be the last. Declining water tables will continue to draw the most attention, but they are not necessarily the most crucial problem. Keeping water in the aquifer will help only so long as the water can be kept clean enough to use.

VII

TOO IMPORTANT TO STOP

NEW AGRICULTURAL PATTERNS are emerging on the High Plains. Cotton—once rarely grown north of Tulia, Texas—is now found as far up the map as southwestern Kansas, where the east–west line of US 160 has become accepted as the practical northern limit for this heat-loving plant. Millet—birdseed—has become a common crop in Colorado. Dairies have moved in almost everywhere. Corn harvests are expanding as farmers plant it not just for human consumption, but for livestock feed, ethanol, and raw materials for plastics. Alfalfa, once rare on the plains, has become a staple.

The thread that runs through all of these changes is an assumption that the Ogallala Aquifer will survive—at least for a while, at least in some form. All the new patterns require access to more water than plains rains can dependably supply. The Ogallala is the only other broadly available water source.

For the most part, High Plains residents understand this. But there is a curious disconnect between what they understand and what they do. They know that the water table is declining, and they know why. They forge ahead anyway, buoyed by the optimistic outlook, common to most human endeavours, that something will turn up—because it always has, because technology can create miracles, but mostly because what they are doing seems too important to stop.

Dairies are a good example. Once uncommon on the High Plains, they have begun sprouting throughout the region like crocuses in mud time. The first person to point this out to me was Oasis State Park's Jim Whary, who complained that the smell from nearby dairies was causing problems for people using the park's campground. "Luckily, the wind rarely blows from the east," he added. "The dairies are all east of us, over around Clovis." He thought they were moving in from California. "Cities over there have been expanding onto dairy land. Farmers sell out for exorbitant amounts, and then they come over here and buy twice the acreage."

"Don't they understand that the water's disappearing?" I asked.

Whary nodded. "Yeah," he opined, "but they're counting on the government to bail them out."

I got a similar—if somewhat softer—take on the issue from Jeff Johnson, a farm manager and doctoral candidate in agricultural economics at Texas Tech University in Lubbock. "What's happening, mostly, is that dairies are coming out of California," he stated. "There isn't water over there anymore, through aquifer decline, or because of water markets or environmental regulations. So they're moving here—one, for the climate, and two, because our environmental regulations are not as strict as California's. And we've got water."

"How long is that going to last?" I asked.

Johnson shrugged. "We ought to have enough for dairies," he said. "Water is going to go to the highest-value crop, and dairies produce a whole lot higher-value crop than cotton. As long as we've got any water at all, I think the dairies will stay."

"Is there going to be enough grain?"

"Oh, yeah. The dairies can haul grain. They can haul grain from the North Pole, if they want, to make dairies viable."

———————

VERY WELL. Dairies *can* haul grain. They can haul water, too, if they can find a place to haul it from. So can cotton farmers and corn growers. The problem is not dairies or cotton or corn, anyway; it is dairies *and* cotton *and* corn. And alfalfa and millet and beef cattle and lawn sprinklers and every other use that demands a piece of the large but limited Ogallala supply. Individually, there ought to be enough water for any of them. Collectively, they are going to run out, and each of them is going to demand that all of the others have to run out first.

And so we come to the most important issue facing Ogallala users today: divvying up the pie. Water is a scarce resource on the High Plains. How is it to be allocated? Who decides who gets how much? There are established answers to these questions, but as the supply continues to shrink, the established answers have begun to raise as many problems as they solve.

VIII

PAPER WATER

I F THERE IS A FLATTER town than Hobbs, New Mexico, any-where in the world, I don't think I want to see it. Pasted to the sur-face of the Llano Estacado thirty miles in from its southwestern edge, the city is completely—one might say, aggressively—horizontal. The theme continues out of town, across farm fields, oil fields, and airfields, 360 degrees of level landscape as far as the eye can see. Level, and dry. New Mexico is the driest state in the Union, and Hobbs is one of the driest places in New Mexico. Rainfall is a scant twelve inches per year; surface streams do not exist. There is oil under Hobbs, and that is why the town was built. But it couldn't survive without the Ogallala Aquifer.

Happily, the aquifer is there, at least for now. Hobbs sits over the Lea County Groundwater Basin, one of the few places along the west-ern rim of the Ogallala where the aquifer's contents did not drain out after it was cut off from the Rockies. Saturated thickness is between one and two hundred feet throughout most of the forty-mile-wide basin. The situation is encouraging—and misleading. As thick as it seems today, the water was once seventy to eighty feet thicker. The Lea County Basin is draining. Nature could not accomplish that task in three thousand millennia. Wells have managed to get nearly a third of it done in less than a century.

But wells are not really the problem. Wells are tools, and like all tools, they will only do what they are wielded to do. Water use is facilitated by wells and pumps, but it is controlled by policies and laws. And in Hobbs, as elsewhere on the High Plains, the policies and laws have not quite caught up with what the wells and the pumps can do.

"WATER WAS GOING TO last forever, just like oil," says Dennis Holmberg. We are sitting in a Howard Johnson's on the eastern outskirts of Hobbs, drinking coffee that, like everything else wet around here, is made from Ogallala water. Holmberg is the county manager for Lea County. His round, bespectacled face wears the bemused expression of one who knows that the human race is capable of both inspired greatness and overwhelming stupidity and is never quite sure which one will surface next.

"You know," he observes, "this was truly the last frontier. This was the Wild West. Most of the communities here didn't exist until about 1925, when they struck oil. And the oil and gas companies—as good as they maybe have been to us—they have had free reign. So we used water. We used pure, pristine water in flood projects, injecting it into the ground to force the oil out. We did a lot of furrow irrigation. We did the worst things you can do. We just thought it was going to flow forever."

As long as water was going to flow forever, nobody gave much thought to it beyond the nuisance involved in getting it out of the ground. It was when they discovered it wouldn't flow forever that things got interesting. As the water table began to drop, the people of Hobbs discovered that state law—New Mexico law, the law of the driest state in the Union—wasn't going to stop the decline. In some cases, it was actually going to allow the water to drain faster.

Things came to a head in 1997, in the spring, as Dennis Holmberg began his third year at the helm of Lea County. That was when the New Mexico Interstate Stream Commission announced a plan to take 650 million gallons per year out of the county's rapidly diminishing water supply and dump it in the Pecos River.

The ISC's scheme grew out of a lawsuit Texas had won against New Mexico nine years before.Upstream water withdrawals had been taking most of the annual flow out of the Pecos before it crossed from New Mexico into Texas, and the U.S. Supreme Court had decreed that the state had to put ten thousand acre-feet of it back. New Mexico has been trying to come up with that water ever since. Lea County's water, which would be obtained by purchasing the water rights belonging to a potash-mining company called IMC Kalium, would take the state a long way toward its goal. It would also leave IMC Kalium high and dry. *They* responded by filing for rights to six thousand more acre-feet per year from the county's rapidly shrinking portion of the Ogallala Aquifer.

The county sued, seeking to have the IMC Kalium deal overturned. The suit was immediately thrown out of court. The judge ruled that the county didn't own any water rights, so they had no standing to sue, and anyway, what was the problem? There was plenty of unappropriated water in Lea County, water that no one had claimed the right to use. That puzzled county officials until they checked the records in the state engineer's office. Those records showed the county with 55,000 acre-feet of unappropriated water. The problem was that the water didn't actually exist.

"You get some engineers out here in the late nineteenth century," sighs Holmberg, "and they drill a few wells, and they say, 'This section has eight hundred acre-feet of water. We'll just say every section has the same amount.' Which isn't true. So you end up with a lot of paper water where you can't get windmill water out of the ground." The courts, and the state engineer's office, were looking at the paper water. Lea County had to find a way to shift their attention to the real stuff.

THE LEGAL STRUCTURES that control access to water are known as water-rights laws. They have been around a long, long time with regard to surface water; the Code of Hammurabi included a couple of them as early as 1790 B.C. With regard to groundwater, regulation is much more recent. Water beneath the earth was neither well enough understood nor heavily enough used to attract lawmakers' attention before the era of the center-pivot sprinkler. Most states over the Ogallala didn't enact any meaningful groundwater-rights legislation until panic over declining water tables began in the 1970s.

New Mexico's Underground Water Law was passed in 1931. That put the state well ahead of the curve, but it also put it out there without much in the way of models to follow. The model, by necessity, became surface-water law. That had advantages and disadvantages. The principal advantage was that the law could follow standard and well-tested formulas. The principal disadvantage was that many of those standard and well-tested formulas didn't fit.

Water-rights laws must solve three problems. They must declare who owns the water, decide who has the right to use the water, and specify how much of the water each user may take. These are closely related but distinct questions, and each state answers them in its own way.

The question of ownership relates to the reservoir; that is, to the water before it has been removed from a stream or pumped from a well. Is it a public resource, or does it belong to private individuals? Is it a separate legal entity, or is it legally attached to the land on which it is found? With regard to groundwater, all High Plains states except Texas and Oklahoma have declared it to be public property. In Oklahoma, it belongs to the owner of the overlying land. In Texas, it belongs to no one until it is pumped out of the ground—an anomaly about which I shall have more to say later.

The right to use water is separate from the question of who owns it. In legal terms, water is transient, transferable, and severable: transient because it moves from place to place on its own, transferable because it can be picked up and moved by humans, and severable because there is no physical reason why any part of it cannot be separated from the rest. (This last is trickier than it sounds. A lake is not severable, because what you do to your part of it will affect your neighbor down the shore. The water in the lake *is* severable, because you can dip a cup into it and take some out. Taking the cup may bother your neighbors; what you do with the water after you take it may not affect them at all.) Because water is transient, transferable, and severable, what applies while it is in a reservoir does not necessarily apply while it is in use. Whether or not you own a reservoir is one thing; whether or not you are allowed to pump water out of it is something else altogether.

All High Plains states require those who take groundwater to put it to beneficial use. You cannot simply pump it out of the ground for the hell of it, even if—as in Texas—nobody owns it. Beneficial use provides a sort of rough first-stage filter for water users: Only those whose proposed use meets their state's beneficial-use definition will be allowed to pump water at all. What qualifies as "beneficial" varies from place to place, but the definition in all states includes water for homes and livestock, water for irrigation, and water for industry, and many add intangibles, such as recreation and environmental maintenance, to the list.

Beyond beneficial-use requirements, restrictions on who may use groundwater vary widely. The three states along the western rim of the aquifer—Wyoming, Colorado, and New Mexico—require permits for all wells, including those for domestic water supply and drinking water for livestock; the other five states require permits only for high-capacity wells, such as those used for irrigation and municipal water systems. Permits may be rejected for proximity (wells may be required to be a certain distance apart) or for overappropriation (portions of an aquifer may be judged to be supporting as many wells as they can). Colorado, New Mexico, and Oklahoma allow the owner of a groundwater right to sell that right to someone else; the other five states tie rights to individual wellheads, although most will allow the owner of a well to sell the water itself once it has been pumped out of the ground.

The hardest water-rights issues to resolve are those that determine how much water can be pumped. These are difficult for two reasons: they only arise when battles caused by water scarcity are already brewing, and they must deal with many, many variables, most of them situational in nature. Water law is not made in a vacuum; it is made by real people dealing with real, constantly shifting requirements, and as a result it is always changing, never quite adequate, and very, very messy.

There are two principal approaches to pumping restrictions: *prior appropriation* and *correlative rights*. Prior appropriation uses the date that a right to pump water was granted by the state to determine priorities among well owners; the newest rights (called "junior rights") are the first to be cut off if a shortage develops. Correlative rights treat all users equally in times of shortage, cutting everyone off from a percentage of their water rather than giving all to some and none to others. Those, at least, are the two forms of rights that occupy the theoretical world. In the real world, neither pure prior appropriation nor pure correlative rights actually exists. All states, whatever they call their systems of rights, use a combination of both.

Take, for example, well-spacing rules. These rules attempt to prevent wells from influencing one another by placing them far enough apart that their cones of depression—the cone-shaped holes in the water table, centered on wellheads, that develop during heavy pumping—will not overlap. Since they apply to all landowners equally, well-spacing rules clearly qualify as correlative rights (spacing is blind to ownership; the rules are the same whether the proposed wells are on one person's property or on several). All High Plains states have some form of well-spacing rules, at least for high-capacity wells in regions of water table decline.

Or take moratoriums on well permits. These apply to everyone in the affected region, so they may appear to be correlative, but they are actually pure prior appropriation. The junior rights get cut off in the most extreme manner possible, by never being allowed in the first place. Most Ogallala states have had moratoriums of one form or another in place over at least part of their share of the aquifer for at least part of the past thirty years.

A particularly complex case is presented by restrictions on draw-down rates. These can take several forms. South Dakota is the extreme case: It allows no depletion of the aquifer at all. (Actually, this rule is almost irrelevant; it does not apply to Indian lands, which include nearly all of the Ogallala's land surface in the Mount Rushmore State.) The other seven states allow limited amounts of depletion to take place, but they vary widely in the manner by which depletion is monitored and controlled. Some impose caps on the amount of water that may be withdrawn; others have caps on the speed of water table decline, which are more flexible than single-well pumping limits but considerably harder to police. Some mandate broad draw-down limits throughout a region, while others tie limits to the area directly around each well. The time period for withdrawal restrictions may be annual (1.4 acre-feet of water per acre of land per year in the Oklahoma Panhandle) or long-term (40 percent of the aquifer's saturated thickness over twenty-five years in southwest Kansas). Whatever their form, draw-down limits are correlative in intent: the problem they set out to solve is recognized as regional in scope, and the limits are meant to apply equally to all. But the rules through which the limits are applied, in most cases, are a form of prior appropriation. Some states cut off junior rights to control draw-down rates; others create de facto moratoriums on new permits by requiring them to fit within draw-down caps that have already been reached. Only in Nebraska and Oklahoma are draw-down limits correlative in application as well as in theory. Nebraska reduces permitted amounts by a uniform percentage in times of drought; Oklahoma imposes strict upper limits on all wells at all times, reevaluating the limits every few years and adjusting them as necessary. For everybody.

LITTLE OF THIS COMPLEX web of legal paraphernalia was in place in 1931 when New Mexico passed its Underground Water Law, so the legislators were forced to fall back, for their models, on surface-water concepts. Just how far they fell back may be seen in the law's jurisdictional language, which declares the state's ownership of all "underground streams, channels, artesian basins, reservoirs or lakes, having reasonably ascertainable boundaries." That was language no knowledgeable person would use for groundwater, even in the 1930s, but it fell comfortably into the structure of the prior-appropriation doctrine already in use for the state's rivers and lakes, and that was what the legislature wanted it to do. For the most part, the bill's authors merely rephrased surface-water law, but there was one exception. The exception had to do with what the law called "mined basins," the areas in which groundwater withdrawals significantly exceeded groundwater recharge. A mechanism was provided to establish these basins as legal entities: once established, they were under the jurisdiction of the state engineer, who could formulate special rules for them. Lea County lost no time in taking advantage of this part of the law. The Lea County Groundwater Basin was officially designated in 1931, the same year the Underground Water Law was passed. The state engineer declared a cap on permits for new wells punched into the basin, but there was a catch: The cap was calculated on the paper water, and it was as much as fifty-five thousand acre-feet too high.

LEA COUNTY TRIED to document the discrepancy between the water they had on paper and the water they had in the ground. They immediately ran into problems. "We have a spaghetti-like network of pipes draining water out of here," complains Holmberg. "The state engineer's office admitted they had no idea how much water was going out of the aquifer." Only part of that dearth of data was due to the complexity of the problem; the rest was self-imposed. "They decided it wasn't important to maintain drilling records," Holmberg explains, controlling himself with an obvious effort. "You know, here we are, with a finite amount of water—we're discussing piping it someplace else—and they stop—" His eyes squeeze shut. "I have to think it was deliberate," he continues after a moment, a little more calmly. "They didn't want to show what the declines were. So we pay to have the U.S. Geological Survey continue that work, monitoring wells, because it's valuable. I mean, it's forty years of records, and if they're not going to maintain it—" He lets the sentence trail off. It is obvious what he is thinking. Continuity is necessary to establish trend. Failure to continue monitoring would create an island of data: interesting, but of limited usefulness for long-term planning. And long-term planning was what Lea County was going to need to fend off attacks on their water, such as that represented by the Interstate Stream Commission's Pecos River enhancement scheme.

The county's first act was to organize. The Lea County Water Users Association held its initial meeting in a conference room in the county's Cultural Center building on December 15, 1997. Mindful of the need to establish both legal standing and the legal authority to act, the organization designed itself carefully. Membership was limited to what Holmberg calls "entities with the right to enact and enforce"— federal, state, and local agencies, including the county government, city governments, and the federal Natural Resources Conservation Service (the former Soil Conservation Service). The Water Users Association might not be able to create enforceable rules, but any rules the members agreed to would get created and be enforced through the individual members' own regulatory systems.

The next step was to create a planning document. New Mexico, in common with most states in the West, has a state water plan that is regularly updated, but state law also allows for regional water planning. Demonstrating that a diversion scheme such as that planned for Pecos River enhancement was in violation of a regional plan might not stop the diversion, but it would certainly be a powerful argument against it.

"So we began the plan," states Holmberg. "But we also had to do something of a legal nature to stop further inroads on our water while the plan was being completed. We had, on paper, fifty-five thousand acre-feet of unappropriated water rights. When we started to complain about our water being taken away, the response from the state engineer's office was, 'You've got fifty-five thousand acre-feet you're not even using. What do you need?' So we began, then, to file. We filed on all of those fifty-five thousand acre-feet. It's the largest water-rights filing in the history of the State of New Mexico." It is also, quite possibly, the weirdest. Where else will you find a series of water-rights applications filed for the sole purpose of using up water that doesn't exist?

Weird or not, the filing has done its intended job. There is no water surplus in Lea County anymore—either real or on paper—and the water buzzards have stopped circling. The Interstate Stream Commission's bid to purchase IMC Kalium's rights to Ogallala water has quietly gone away; so has the potash company's request for rights to more water for its own use. The county's water plan has been completed, filed, and certified by the state engineer. For the moment, Lea County seems in control of its own water destiny.

Whether or not the county remains in control will depend to a large extent on what happens across the border in Texas, whose peculiar water laws may eventually render moot anything the Lea County Water Users Association can accomplish. What passes for water-rights doctrine in the Lone Star State is something called the Rule of Capture. In essence, anything you can pump out of the ground is legally yours.

Nobody owns groundwater in Texas. It is considered a free good, like air or sunshine—available to anyone. This concept stems from case law rather than from legislation; it originated in a 1904 Texas

Supreme Court decision known as *Houston & T.C. Ry. Co. v. East.* Texas courts have upheld it ever since, most recently in 1998, and the legislature has not been inclined to challenge it. A few qualifiers have been added over the years—water must be used for beneficial purposes, and local regions may, with the consent of a majority of voters, enact some limitations on wells that pump more than twenty-five thousand gallons per day—but property rights to water do not exist until the water has been brought to the surface. If your neighbor's well drains water from beneath your land, you can do nothing about it; your neighbor has as much right to that water as you do.

In an article published in 1967, ecologist Garrett Hardin pointed to what he called "the tragedy of the commons"—the tendency to overuse and abuse common property out of a fear that someone else will get your share before you can. The Rule of Capture is a textbook example of the tragedy of the commons in operation. Everyone who uses water from an aquifer such as the Ogallala, where recharge is minimal, will cause a drop in the water table; as an individual, you have no choice to preserve it, only a choice to pump it yourself or to watch someone else pump it out from underneath you.

For residents of Lea County, where the Rule of Capture does not hold, the fact that Texas continues to cling to it is particularly galling. Lea County residents must abide by New Mexico's regulations to withdraw water from the ground. Wells just across the border have no such restrictions. A well in Texas can draw all the water out from under a piece of property in New Mexico, and there is literally nothing that the property owner can do about it, except watch. The state will neither intervene nor, usually, grant a permit allowing the New Mexico resident to pump the water first.

"The State of New Mexico has not been wanting to deal with Texas on water management," complains Holmberg. "Well, maybe that's not important if you're in Albuquerque. But our biggest problem here on the Ogallala has been that there are no water restrictions over the border in Texas. The restriction has been how big a hole can you drill, and how big a pipe can you put in the ground, and just go for it. We have a different way of looking at that." The Water Users Association has been to the New Mexico legislature with what Holmberg calls a rather "out there" idea to create a five-mile-wide no-drilling-restrictions zone along the Texas border. "Our phrase was, if we're going to have a race to the bottom of the barrel, let's try to win it," he smiles. "Let's not put restrictions on our side." So far, the chances for actual enactment of such a zone have ranged from nil to zero, but at least the county has called attention to the problem.

The better solution, of course, would be to create some restrictions in Texas. Lea County is working on that, too. "We go over and meet with people in Lubbock," says Holmberg, "because we don't need a water plan for New Mexico that says this, and a water plan for Texas that says that. We've got to get rid of the political boundaries." As long as water-rights regulations change at borders, borders are going to be a problem. There is only one Ogallala. It contains great variation, but the variation follows its own design; it has no regard for irrelevant human constructs, such as states and counties and private property lines. If we are to preserve the opportunities that this aquifer provides, we are going to have to learn to manage it on its terms, not on ours. Management on the aquifer's terms requires a holistic approach. The states' separate takes on groundwater law make this a difficult task.

But there is another possibility. Instead of eliminating the boundaries that prevent us from managing the aquifer as a whole, we can make them irrelevant. To do this, it is necessary to look small, not large—to aquifer subunits, and to a form of governance that allows rules to be tailored independently to the characteristics and needs of each individual area. The Lea County Water Users Association is a step in that direction. Further steps have been taken elsewhere.

IX

TO DETERMINE THEIR DESTINY

W HEN RAIN FINALLY COMES to the High Plains, it does not mess around. I found that out one cool April Wednesday as I drove across Kansas from Great Bend to Garden City. Sheets of water poured from black clouds; the windshield wipers struggled to maintain a clear space in the rivers rushing down the glass. Thunder roared like a chorus of angry bowlers. Getting out of the car felt like stepping, fully clothed, into a cold shower.

The rain departed abruptly in midafternoon, gusting off toward Kansas City on the back of a whirling, Wizard-of-Oz wind. Its receding edge was a wall across the prairie, thousands of feet high, at least a hundred miles long, and as straight as if it had been sliced by a paper cutter. In its wake came blue sky and breathing earth reveling in its recent bath. I checked into a Garden City motel and went to find the Arkansas River. Except for a bit of dampness on the upper surface of its gravel, the bed of the river was bone-dry.

Like Hobbs, Garden City could not exist without the Ogallala Aquifer. It has no other dependable source of water. The river has been gone since before 1900; the rains, though furious, are fleeting and uncommon. The underground ocean remains, at least for the present. To stretch that present as far as possible, Garden City relies on a blonde, unassuming man named Hank Hansen, and on a legal tool of a type that is as common on the High Plains as it is rare everywhere else: a groundwater management district.

I called on Hansen later that afternoon. The Southwest Kansas Groundwater Management District occupies a small suite of offices in a little strip mall on the southeastern edge of Garden City, near the local community college; the dry ghost of the river lurks barely a mile away, beyond the tracks of the Burlington Northern Railway. The district's literature identifies it as "GMD#3." It was the third of Kansas's five groundwater management districts to be formed, and remains the largest. From Garden City it stretches east to Dodge City, south to Oklahoma, and west all the way to Colorado—twelve counties in all. As district manager, Hansen oversees a staff of five, 10,500 irrigation wells, and an annual budget of nearly one million dollars.

"It was at the request of local people that this district was established," Hansen told me, "and I get the firm impression that the reason was to have some sort of local control mechanism put in place to address the rate of aquifer decline. So this office began, twenty-five years ago, to establish some really serious, locally driven control measures." He ticked off some of them. Well-spacing criteria. Aquifer-depletion criteria. Criteria for closing parts of the district to further appropriation. Mandated water meters. "Those were all things that were inspired locally, that local people put in place through this office."

The word "local" appears often in Hansen's discourse for a reason: An emphasis on local control is crucial to the groundwater management district concept. Restrictions necessary to slow the decline of an aquifer can seem draconian. Imposed from above, they would foment rebellion. Imposed from within, the same regulations can be put in place with scarcely a murmer.

GMD#3's rules governing permits for new irrigation wells provide a good example. These are handled on a township-by-township basis.

(The word "township," as used by westerners, means a surveyed square of land six miles on a side; it may or may not have an actual town on it.) If the part of the aquifer within a township has been depleted by 20 percent or more, or if its saturated thickness is less than fifty feet, then that township is closed; you cannot get a permit for a new well. If these two criteria have not been exceeded, you will be allowed to apply for a permit, but in order to actually get it, you will have to certify that the aquifer will not be drawn down more than 40 percent in twenty-five years within a two-mile radius of the place you intend to drill. "We have 250 townships in this district—it's a large district," Hansen observed. "Of those 250, we have about forty-six, at this moment, that are technically open to appropriation. Applicants still have to comply with the well spacing and the two-mile radius. It's fairly difficult, in most of what we call 'open' townships, to find an area where you can still drill your well." When a similar set of restrictions on farmers' surface-water rights was imposed in 2003 by the U.S. Interior Department in Oregon's Klamath Basin, the farmers dynamited the irrigation gates. In Kansas, through their elected representatives on the board of the groundwater district, the farmers have imposed the restrictions on themselves.

Non-farmers have not always understood. Hansen recalled hearings in 2002 regarding a district-wide moratorium on new well permits that the board considered but ultimately rejected. "As we went through that process," he told me, "numerous groups came to testify against closing the district—mostly groups, in my impression, representing economic development. It wasn't farming groups who came to protest; it was urban people worried about potential slowdowns in the economy that might be caused by closing the district to appropriations." Those concerned about the economy could not grasp the fact that the water they were demanding farmers have access to was no longer there. The farmers could, because their own wells, and their own groundwater management district, were telling them so.

HIGH PLAINS GROUNDWATER management districts first appeared in Texas, whose Underground Water Conservation Districts Act passed the state legislature on June 2, 1949. The Lone Star State's unique approach to underground water rights might make it seem an odd leader in the groundwater management field, but in fact, the Rule of Capture encouraged this development. The Rule of Capture has always been about liability, not about management. It prohibits you from suing your neighbor for draining water from beneath your land. It does not prevent you and your neighbor from getting together to jointly manage your water. You might even say it actively encourages joint management. Without an agreement in place, both of you will have an incentive to grab your share before the other guy does—the race to the bottom of the barrel that everyone tries to win. A management agreement that treats all parties equally removes that incentive. And a management agreement—writ large—is all that a groundwater management district really is. You still cannot sue your neighbor for depleting your well in Texas, but you can create rules to control well depletion that you and your neighbor are both legally bound to follow.

To make certain that Texas's groundwater management districts were governing with the consent of the governed, the Underground Water Conservation Districts Act stipulated a lengthy process for creating and certifying new districts. As a result, it was not until 1951, two years after the act's passage, that the first district became operational. That was the High Plains Underground Water Conservation District, headquartered in Lubbock and encompassing six counties (it now contains all or parts of fourteen). Two more came on line four years later: the North Plains District just north of the Canadian River, and the Panhandle District just south of it. The next fifty years saw most of the rest of the state follow suit. Today there are eighty-seven groundwater conservation districts in Texas. Eleven of these are over the Ogallala, blanketing the resource almost completely; of the state's forty-three High Plains counties, only six are not included, at least partially, in the boundaries of a groundwater district.

Kansas passed its Groundwater Management District Act in 1972. Like the Texas act on which it appears to have been modeled, the Kansas version spelled out a careful set of controls to make certain that groundwater management districts were created and managed with the consent of the governed. All five districts that exist today were in place by 1979. The three western districts are over the Ogallala; the other two are over what Dan Zehr of GMD#5 in Great Bend calls "reworked Ogallala"—Ogallala gravels that were picked up by rivers and deposited farther east. These deposits are hydrologically connected to the big aquifer and are usually treated as part of it.

Colorado and Wyoming also have groundwater management districts in place, although only Colorado calls them that; in Wyoming they are called Groundwater Control Areas and are created, not by popular vote, but by a decree of the state engineer (their governance remains local). Nebraska's Natural Resource Districts have powers over groundwater that parallel those of groundwater management districts in other states. Oklahoma has Groundwater Regions, which have some of the same functions as groundwater management districts; New Mexico has Groundwater Basins, and quasi-official bodies like the Lea County Water Users Association to take care of them. Of the eight High Plains states, only South Dakota continues to hold determinedly to a statewide, top-down model of groundwater management, and there it really doesn't matter, because most of the Ogallala within the state is under the control of the Lakota Nation.

IN GARDEN CITY, Hank Hansen explains how the Kansas version of the bottom-up model works, using his own Southwest Kansas Groundwater Management District as an example. Local control is paramount, beginning with the six-step procedure by which a groundwater management district is founded. A district must:

1. File a declaration of intent with the office of the chief engineer of Kansas,
2. Be investigated and certified by the chief engineer,
3. File a petition for an election,
4. Hold the election and have the results certified,
5. File for incorporation with the office of the secretary of state and, following incorporation,
6. Hold a meeting to elect a board and formally organize the district.

All steps except the second take local initiative, and three of the five require an overt demonstration of support from the community. There was plenty of that in southwest Kansas; the election to form the district, held on February 24, 1976, resulted in a vote of 1,155 in favor to 230 against. Incorporation came a month later, on March 23. On April 6, capping a process that had begun almost three years earlier, Southwest Kansas Groundwater Management District #3 held its first meeting. Delays had almost caused it to become District #4: The Northwest Kansas GMD, which had begun the process nearly a year later, was by that time just over a month behind. It held its first meeting in Colby on May 24.

The theme of local control carries through the districts' operating rules as well as their means of creation. Groundwater management districts in Kansas are governed by an elected board, with one member from each of the district's counties plus three at-large members representing specific groups of non-groundwater-irrigating stakeholders: municipal water users, industrial water users, and surface-water users. Funding comes from tax levies; in GMD#3 there are two of these, one on irrigated land and the other on pumped water. Each district makes its own rules, subject to approval by the state chief engineer; these rules then become part of the Kansas Administrative Regulations, the state's body of administrative law. That makes them legally enforceable within the district, but because they are created within the district—rather than imposed from without—enforcement is rarely necessary. Voluntary compliance is almost universal.

"If people participate in the establishment of rules, they demand enforcement of those rules," Hansen pointed out. "I'll give you a quick example. We do over two thousand well inspections a year in this district. We processed maybe two hundred cases of noncompliance last year. That's nothing out of more than two thousand inspections. And almost all of those were settled immediately." He smiled. "The inspector leaves a little orange tag. We find out about most tags from the operators before the inspector gets back to the office. The operators want to know how to fix the problem, not contest the inspection. We had one case last year—one case out of more than two thousand— where the well owner didn't want to comply."

LOCAL CONTROL DOES not always lead in the direction that water managers might prefer. In the Northwest Kansas Groundwater Management District, the staff and the board spent many hours crafting a complex, delicately balanced plan to achieve zero depletion—no net lowering of the water table—only to see it shot down by the district's voters. The Great Bend District has had to abandon a similar scheme, called in its case "sustainable yield." In Texas, the Panhandle District has instituted water metering "with a little reluctance, honestly, on the farmers' part, because that smacks of regulation"—even though water meters help the farmer manage his own water use better, for very little money (the district picks up half the cost of installing the meters).

In Wyoming, two of the three Groundwater Control Areas over the Ogallala are largely closed to new wells, while the third remains wide open. The state would prefer to see all three closed, but finds it prudent to accede to local wishes. "That's just what they want to do," irrigation specialist John Harju told me in the state engineer's office. "We take their advice, and we act on that advice." Whether the advice actually serves the best interests of those who are providing it is an inquiry it seems prudent not to pursue.

Despite problems such as these, most people involved in High Plains groundwater management feel strongly that local control should be retained—and worry that it will not be. The storm that ushered me into Garden City was nothing compared to the storm that will erupt on the High Plains if the states—or worse, federal agencies—attempt to substitute outside fiats for homegrown solutions. All but a few High Plains residents understand that it is necessary to create regulations to extend the life of the Ogallala. What they oppose—what they fear—is that outsiders will attempt to create the regulations for them.

"From my viewpoint, in the trenches here, we're having a good deal of positive impact," Hank Hansen observes. "But if we're going to all of a sudden have a package of incredibly restrictive laws, how is enforcement going to take place? We struggle with budget priorities now. If we're going to cut people's rights back, how are we supposed to police that?"

Wayne Bossert, Hansen's counterpart at the Northwest Kansas Groundwater Management District and the chief architect of that district's rejected zero depletion policy, agrees with Hansen. "I could argue that we all have an ax over our heads, because the legislature's getting pretty damned impatient," Bossert told me in his Colby office. "Local control is one way to approach groundwater management, but it isn't the end-all, be-all. We've got to produce, or else the legislature's going to say it's a failed experiment. And if the state takes it over—" He shook his head. "They don't have the manpower or the budget to do it. Especially now. They don't have enough budget to open their doors in the morning, hardly, let alone get as sophisticated as we have been in the management and the monitoring and the enforcement of groundwater regulations. If they take it over, it's going to set us back a long, long time."

TWO

THE GHOSTS

OF ANCIENT

RIVERS

X

PEERING AT THE EDGE

THE TOWN OF OGALLALA, Nebraska, nestles compactly on the north bank of the South Platte River a few miles downstream from Emigrant Crossing, where the pioneers' wagons finally stopped following the river and struck out cross-country on their way to South Pass and distant Oregon. This is where the founders of Colorado reared up and took a large bite out of southwestern Nebraska, creating the Nebraska Panhandle; Ogallala just barely avoided becoming Ogallala, Colorado. In the 1870s, Ogallala was the northern terminus of the Cimarron Trail, over which Texas cattle were driven to the railroad for shipment to eastern stockyards. The town was wild, woolly, and wide open, which may explain how it came to be named for the people then known as the Ogallala Sioux. The Sioux did not often come to that region—the South Platte was Arapahoe territory—but they certainly had a reputation for being wild. Perhaps this was because whites insisted on calling them "Sioux," an abbreviated form of the Ojibwa *nadouessioux*, meaning "wretched little foreigner" or "small rattlesnake." The Sioux called themselves "Lakota," which means "friend." The Ogallala—or, to use their own preferred spelling, Oglala—were, and are, the most numerous Lakota band. The word *oglala* is usually translated as "to scatter one's own," but it was also the name of a faintly contemptuous action—a flicking of the wrist, palm down, while opening the fingers—that was part of Plains sign language. It appears to have been that action, rather than its literal meaning, that tied the word to the people. Ogallala is possibly the only county seat in America that was named for a rude hand gesture.

The Platte River Breaks begin at Tenth Street and rise quickly from the flat river bottom to the even flatter High Plains, two hundred feet above. The breaks are cut by draws. The walls of the draws reveal their underlying structure: a thick deposit of pebbles and cobbles, pinkish-gray in color and piled in no particular order. That is the Ogallala Formation, and this is its type locality—the place where it was first scientifically described. The year was 1896; the scientific describer was a young New York geologist whose name clearly indicates that his family expected him to do something entirely different with his life.

Nelson Horatio Darton came to geology by a roundabout path. His father, a Long Island shipyard owner, had destined him for a career at sea; Nelson confounded that by leaving school at the age of thirteen and apprenticing himself to his uncle in Manhattan, a pharmacist, where he quickly became inspired less by pharmaceuticals than by his uncle's rock collection. He began to read geology books and to do independent research, especially on groundwater. A catalog he prepared of articles about the geology of the state of New York came to the attention of G. K. Gilbert of the fledgling United States Geological Survey, who asked the young autodidact to prepare a similar catalog for the Appalachian Mountains. That led to a research appointment with the USGS and, a few years later, to a similar position with the New York Geological Survey, where Darton quickly became known as an authority on artesian wells. He was not yet thirty years old.

In 1894, the USGS asked Frederick Newell, the chief of its hydrographic branch, to put together a team to evaluate groundwater reserves in the western United States. Newell thought immediately of

Gilbert's young protégé in New York, and thus it was that, in the spring of 1896, N. H. Darton (as he now signed himself) stepped off a train in Ogallala, hired a carriage, and drove east along the South Platte River, looking for a place where he could examine the underlying structure of the bluffs. The place he chose was three miles from town, back a bit from the river, near the mouth of an unnamed draw. Darton noted the characteristics of the formation—pinkish-gray pebbles and cobbles mixed with sand and numerous small fossilized plant inclusions, Tertiary Age—and named it, according to established custom, after the nearby town. He was still wearing the suit and tie he had put on for the train, a field outfit that was to become a Darton trademark.

Contemporary photographs of the geologist show a slender and aristocratically boned face, a slightly receding hairline, a humorless mouth beneath a dark, full moustache, and intense, piercing eyes. Colleagues considered him a pain in the ass. One recalled how he had been examining an outcrop along a trail in a remote region when Darton came strolling around the corner. "I've made my guess about what this is," the other geologist remarked. "What do you think?"

Darton regarded him coldly. "I never guess," he replied. "I study the facts, and I know."

Darton was not a popular man at parties, but his work was revered. His meticulous, beautifully rendered maps were works of art as well as of science; his pioneering experiments with the camera as a field tool not only induced other geologists to follow him but produced a body of work that put him in the first rank of early landscape photographers. He was also inhumanly accurate. A well driller near Edgemont, South Dakota, once asked the great groundwater expert how deep he would have to go to find water. Darton told him three thousand feet. The skeptical driller drilled anyway—and drilled, and drilled, and drilled. He hit water at 2,965.

Given that sort of prodigious attention to detail and innate sense of the land, it should come as no surprise to find Darton, two decades after his visit to Ogallala, coming across an outcrop of pinkish-gray pebbles and cobbles near Tucumcari, New Mexico, and recognizing that they were not just similar to the ones he had described back in Nebraska but were, in fact, part of the same formation. A humongous sheet of erosional debris spread across eight states, split now by rivers but originally continuous. Many source materials, one history. The name is an accident of geography, and means nothing. It is the pebbles and cobbles, and the ghosts of ancient rivers among them, that count.

PEBBLES AND COBBLES are the theme. There are nearly infinite variations. There are variations in mineral content and particle size; there are variations in the thickness of the formation and in its depth below the surface of the ground. Parts of the Ogallala are deposits of nearly pure sand, others are gravel, still others are boulders, some the size of small houses. In some places ancient stream channels may be traced; in others, channels are absent and the materials are stacked randomly upon one another. These hodgepodge piles, which give geologists fits, were laid down as overbank deposits in times of flood. Horizontal layers of caliche—a weak, grayish-white stone with a high calcium content—lace through the aquifer in the south. The layers were formed by minerals left behind on the ground when the water carrying them evaporated away, and thus are indications of old surface locations. They range in thickness from less than an inch to more than twenty feet. Caliche fractures easily, but it resists erosion. The caprock of Texas's Caprock Escarpment is a massive caliche bed; its protective presence is responsible for the spectacular landforms of the Texas Canyonlands, where the High Plains suddenly break off into a vertical wilderness of cliffs and castles, pillars and precipices, a thousand feet above the rolling landscape below the rim.

The High Plains seem vast today, but they were once much vaster. They are a remnant surface, the core of a prehistoric superplain that once extended many miles further east, perhaps as far as the Mississippi River. Since deposition ended, the edges have been nibbled away. That is the reason for the rise as you enter them: It is an erosion scarp. Streams to the east have been busily carrying away the toe of the deposit; streams to the west have cut north and south along the base of the Rockies, carving the aquifer away from its source. The High Plains are what remain. If you want to study their structure, you have two choices. One is to go to an edge, as Darton did, and examine what erosion has randomly revealed. The other is to go out in the middle and start digging.

XI

TO DIG A HOLE

THE LARAMIE MOUNTAINS

JON MASON KEEPS both hands firmly on the wheel, steadying the red USGS Jeep Cherokee against the wind that is threatening to blow it off Wyoming Highway 211. Behind the Cherokee, the western edge of Cheyenne recedes into the flat infinity of the plains; ahead, the smooth pink knobs of the Laramie Mountains lift into a slate-blue morning sky just beginning to slip toward gray. To the left, a long line of land slants gently upward across the head of the Lodgepole Valley. That is the Gangplank, the one tiny spot along the mountains' thousand-mile front where their connection to the Ogallala Aquifer remains intact.

"There are still a lot of mysteries about this whole Rocky Mountain region, and what has happened, and why the landscape looks the way it does today," Mason says quietly, keeping his eyes on the blacktop. "I'll give you just a hint of one of them, so you can think about it while we're driving. There are mountains west of Laramie called the Snowy Range. They're a lot higher than these mountains we're looking at. The highest peaks here are around nine thousand feet; the Snowy Range tops out at over twelve thousand. And then there's the big Laramie Basin between the two. There are rocks from the Snowy Range in downtown Cheyenne. How did those rocks get there? How did they get across the Laramie Mountains? They're old rocks, but they were transported fairly recently in geologic time. They're just lying there on the surface by the Holiday Inn. That's Ogallala Formation there, so—"

"So they're on top of the Ogallala."

"Right. How did that happen?"

Jon Mason is a large, loosely made young man with a thatch of dark blonde hair and a master's degree in geology from the University of Nebraska at Lincoln. Growing up in York, Nebraska, in the eastern part of the state but still over the Ogallala, he worked in his father's construction business and thought about basements. He wondered why the upper part of the excavations always slumped while the lower part always held a clean vertical line. Living in Ohio a few years later, building B-1 bombers for Rockwell International with a diploma from flight mechanics' school in his bag, he found that he was still thinking about basements. "Eventually I realized I needed to be around other people who like to ask questions about the physical world and think about unusual things," he says today. He moved back to Nebraska and enrolled at UNL. The Cheyenne office of the U.S. Geological Survey hired him fresh out of graduate school in 1995; two years later, the chief of the Wyoming section of the survey's High Plains Aquifer Program abruptly quit, and Mason was offered the job. He has held it ever since.

"Here's another mystery along the same lines," he adds, after a moment. "There are several mountain ranges in Wyoming that have rivers that literally cut them in half. The Wind River is a good example. It starts in the western part of the state, flows east, then turns north and cuts through the Owl Creek Mountains, in the form of Wind River Canyon. Then it goes through the Bighorn Basin, where it's renamed the Bighorn River—same river, it's just got a new name. The Bighorn River cuts through some pretty impressive anticlines, and then it bores through the north end of the Bighorn Mountains in a really spectacular way. So that's part of this mystery. How did that happen? Usually you'd think of a river going around a mountain. How in the world did it cut right through?"

"Are the answers interrelated?" I ask.

"Yes." He points through the windshield to a place just ahead of us where the pavement makes an oblique turn to the right and disappears. "Now we're going down into the place where the High Plains start. They used to go all the way to the mountains, but they've been eroded away. Except for the Gangplank, which is pretty much intact—that's why it's unique." Another brief silence. "A million years is nothing to a geologist, you know. I might have been misleading you into thinking that this mountain cutting, and those rocks by the Holiday Inn, are things that have happened in the last few thousand years. They're not. I'm talking about something that happened over tens of millions of years. It started in the early Tertiary and went on into Ogallala time—the time when the formation was laid down. I like this area down behind the hill, with these hogbacks sticking up. That's a thrust fault. That's the end of the aquifer, right there."

Hogbacks are long, narrow ridges that are steep on both sides. Almost always they are the result of extreme upwarping of what geologists call layer-cake geology: multiple layers of clearly definable sedimentary strata laid neatly on top of one another. The layers tilt upward until they break, forming a fault. Then they tilt some more, lifting one face of the fault out of the ground. The uplifted face forms one side of the hogback; the original ground surface forms the other side. In extreme cases, the high side of the fault may ride over the top of the low side. That is a thrust fault. Thrust faults are barriers to groundwater flow: When water trickling laterally through an aquifer encounters a thrust fault, it usually pools up and spills out at the surface as a line of springs. That, Mason explains, is what is happening here. It is thrust faults along the mountain front, as much as erosion, that have blocked the flow of groundwater and cut the Ogallala off from its ancient recharge areas under the snows of the Rockies.

A few minutes later we are behind one of the hogbacks, on a gravel road edged by willow wands and white columns of budding aspen. The road plunges into a tight little pink-walled canyon beside a loud stream, then climbs past grass and cattle to the point of a ridge coming down from the Laramies. Mason pulls the Cherokee to a stop. "This is a great view of the High Plains," he remarks, gesturing toward the east. "Look how flat that is."

We are on a rounded knoll of brown earth and pink crumbly rock, treeless but spangled with ground-hugging flowers, mostly the familiar pinkish-white blooms of *Claytonia lanceolata*—western spring beauty. The Laramie Mountains rise to the west; the craggy, steep-sided hogback blocks the view to the northeast. The rest of the view is huge: the broad Lodgepole Valley, then the edge of the High Plains, then nothing but flat, flat horizon. The land is green with May.

Mason pulls out a geologic map and spreads it over the hood of the Cherokee. "We're on this gravel road right here. We left Cheyenne— zoom, zoom—" He moves a finger along Highway 211. "Right here is the boundary where the Ogallala disappears. Remember when we were driving, I said, 'now we're driving down off the High Plains'? That would have been here. So this is the White River Formation." The finger moves east. "This is the Ogallala. One way you can tell the Ogallala from the White River is to look at it under a microscope. The White River has a lot more volcanic ash in it. That came from volcanoes farther west, some of them in Idaho.

"This is where we're at right now. This is Horse Creek. Here's where we're standing, up on this Precambrian granite." He taps the map. "This is an area along the mountain front where stuff has been eroded away. It's west of us, but it's actually on the east side of the fault. This granite we're on has been pushed up and over the younger rocks. Here's a cross-section of a well that they drilled, and you can see how they went through granite at the top, and then they drilled down into the younger stuff beneath. That's gotta be exciting to do."

"So this rock we're standing on is Precambrian?" I ask. The Precambrian era is the oldest part of geologic time—the era that forms the foundation for everything else. Little fossilized bacteria—microfossils—are occasionally found in Precambrian rock. Every other living thing on the planet came afterward.

Mason nods. "Yeah. It's the Sherman Formation. It's a granite gneiss, actually. As you can see, it just crumbles." He demonstrates. "This is what's filling up the Platte River right now. It starts out here all sharp and angular, but by eastern Nebraska it's ground down into nice little round sand. That rock over there is the Casper Formation." He indicates the hogback. "It's buried a thousand feet deep or so underneath the western edge of the High Plains. Cheyenne's thinking about drilling some wells into that, in the same well field they're using for the Ogallala—tapping into the Casper Formation as a water source."

I note that, according to the map, the part of the plains we are looking at rests on an angular unconformity—a place where tilted layers of rock have been planed off flat and then covered by later deposits, so that the newer layers lie across the upturned edges of the older ones. Mason nods. "One thing you learn when you study stratigraphy is that only 15 percent of the Earth's history is actually recorded in rocks," he points out. "Eighty-five percent of it is missing. Gone. So I always have to laugh when I watch science shows on television. They show these elaborate histories of how the dinosaurs lived, but really, all we see is this tiny, tiny little window. And you've got to wonder if we're missing the whole big picture, because the important part got eroded away, and what we see is just one little oddball rock that happened to be left behind that doesn't really represent what was going on most of the time."

It would be easy to infer from this that Mason, a geologist, believes that geology is an inexact science. He does not believe that. The details of ancient landscapes may be obscured by time, but the processes that built them are precisely understood. They can be observed directly, because they are the same processes that are active today. The principle of uniformitarianism—the unchanging nature of earth processes through time—is as true now as it was in 1788, when James Hutton published his *Theory of the Earth* and put geology on a sound theoretical basis for the first time. "The present," wrote Hutton, "is the key to the past." If you want to understand how the Ogallala Aquifer was built, look at the Precambrian sands in the Platte River.

Mason is talking about the Laramide Orogeny, the process—seventy million years old and counting—that has been slowly giving us the Rocky Mountains. He hands me a pair of diagrams. One shows level, neatly stacked geologic strata; the other shows what appear to be the same strata, but they are arched upward in the middle, and the top of the arch is gone. "Pre-Laramide, all this was just layer-cake flat," Mason explains. "That's the first diagram. The second one shows the layers beginning to bend, due to compressional forces. Before that happens, the structure is flat, and all the beds are there. But as soon as you start to bend it, the sediments on the top part of the bend begin to erode. As the bending continues and the mountains go up, you erode more and more of those sediments away. In the Laramies—remember, this is Precambrian granite we're standing on—we've eroded away all of the Paleozoic and all of the Mesozoic that used to be above us. At least ten thousand feet of sediments. Gone. And where are those sediments now?" He waves a hand over the big view to the east. "That's where they're at. That's the High Plains. This was the source, and that's where it all is right now."

THE SAND HILLS

A narrow road winds among tall dunes in Arthur County, Nebraska, near the southern edge of the Sand Hills. Leaving the paved Silliasen Highway where it cuts through a corner of John Jensen's ranch, the road crosses a cattle guard, dodges between a marsh and an interdunal lake, and climbs quickly to a low saddle. On the far side is a fence line and a dense little plantation of junipers, and there, to the right and a bit uphill, is the drill rig, just where Jim Goeke said it would be.

The Sand Hills are really sand. Not just sandy dirt, but sand, twenty thousand square miles of sand, blanketing most of western Nebraska and spilling over a dozen miles or so into South Dakota. The "hills" of the region's name are dunes—barchan dunes, linear dunes, crescent dunes, dome-ridge dunes, barchanoid-ridge dunes. Some of the dunes are four hundred feet high and twenty miles long. They have been partially stabilized by grass, but the sod is thin and prone to breaking down; blowouts, where the wind breaches a dune crest and spills naked sand down the leeward face, remain a common occurrence. Geologists call a dune field like this a "sand sea," and this is one of the premier sand seas of the planet. Though you may find larger ones in Arabia, in North Africa, and a few places in Asia, there are no others comparable to it in the Americas. It is odd, but true, that Nebraska is the only place in the Western Hemisphere that belongs in the same geological category as the Sahara Desert.

Goeke ambles over to say hello, shouting to make himself heard over the rattle of the drill rig and the roar of the wind. He is big, rumpled, and jowly, a great unmade bed of a man with mud-stained jeans and a dangling flannel shirttail. "You know Nebraska is the Tree Planter State?" he asks, without preamble. "Arbor Day State."

"I think so," I reply. "Arbor Day started here, didn't it?"

"You know why?" Goeke prods.

"Tell me."

"To stop the damn wind." He snorts. "The wind will make you crazy, right? Once you've been around here a day or two—it doesn't take long—you want some trees or something to slow the damn wind down." He tugs down his scuffed red hard hat and strides back to the drill rig.

Jim Goeke is enamored of drill rigs. The look of a mast against the sky moves him the way another sort of mast might move a sailor. His hands itch for the controls. "The one person who really knows what is going on," he once remarked to me, "is the fellow with his hands on the levers. He's standing on that machine, and he knows how his pump sounds—he knows what the vibrations are—he can tell the rate of penetration, and he can tell by the feel of the levers what's happening down at the bit. He is the one individual who really knows what the hell is going on."

In 1969, Goeke was working toward a master's degree in geology at Colorado State University in Fort Collins. He was trying to model groundwater flow in the basin of Black Squirrel Creek, an intermittent stream for much of its length that rises in the northern outskirts of Colorado Springs and flows east and south onto the plain. Data proved difficult to come by. "I had to make do with subsurface information from drillers' logs. And we twisted it around enough that I got a master's degree out of it. But I realized the whole thing was built on a house of cards, because I had no clue—about the drillers, and the records, and the logs they had kept." While Goeke was struggling with this problem, Richard Nixon decided to invade Cambodia. The Old Main building on the Colorado State University campus became one of the casualties. "The U.S. started bombing," Goeke recalls, "and someone got pissed off and burned down Old Main. My office was in the basement. I was actually sleeping on a cot, in my office, in that building. And they burned it to the ground. Burned every book, every note I ever took. The only thing I had left was a briefcase with the data deck for the Black Squirrel Creek model. So I finished the model and I said 'screw it.' I had credits towards a Ph.D., but I had been totally disenfranchised. And I needed to get a job."

He was offered a position with the Army Corps of Engineers in Florida. He turned it down. A second offer had come through from the Conservation and Survey Division of the State of Nebraska. The work would involve a cooperative test-drilling program run jointly by the state and the United States Geologic Survey. "I jumped at that, because I had been involved in a project where I didn't have subsurface information that I felt was worth a damn. I wanted to be in some sort of a program that would systematically develop good information. I started on the first of July, and on the fifth of July we went up in the Sand Hills and started drilling." He grimaces. "That was a low ebb. I thought that was the most God-forsaken, desolate, barren place in the world. But the more we drilled up there, the more I liked it." It was the beginning of a love affair with the Sand Hills, and with the Ogallala Aquifer beneath them, that has not slackened to the present day.

GOEKE IS CATCHING SAMPLE. At the moment, he is using a purple one-pound Maxwell House coffee can; later, he will employ a six-inch kitchen sieve with a black plastic handle. He holds the can under the outflow pipe of the well. The well is not producing water; what is coming out of the pipe is drilling mud, a thick brown slurry of water and bentonite clay that is circulated constantly through the drill stem to cool and lubricate the cutting edges of the bit and to seal the sides of the hole against collapse. Part of John Jensen's fence has been cut and moved aside, and a recirculating pit, or "mud pit," has been dug in the gap; the mud is pouring into that. From there it is pumped through twin two-inch hoses to the top of the mast, where it reenters the drill stem. Goeke sits on a low white plastic stool, facing the drill rig, holding out his Maxwell House can to intercept the mud as it falls toward the pit. Over and over and over. Each can is emptied into a white five-gallon bucket, where the cuttings the mud has brought to the surface are allowed to settle out. The mud is then decanted into the pit, and the cuttings are carried to a nearby examining table, where they are identified, entered in the well log, and troweled into a small white muslin bag. The drillers stop the drill every five feet and wait while the cuttings come to the surface, fall into the coffee can, and are decanted, poured, identified, logged, and bagged.

This Arthur County well is part of a project called COHYST, an acronym—sort of—for Cooperative Hydrology Study. COHYST is funded by a consortium of several public agencies and a smaller number of private corporations, all of which have an interest in understanding the flow of the Platte River. Because surface water and groundwater are intimately connected, it is necessary to model groundwater flow beneath the Platte's watershed in order to fully understand the river's behavior. Hence, this well. The plan is to drill all the way to the bottom of the Ogallala Formation and into the impermeable Brule Formation beneath. The Brule, a brown siltstone, is the upper member of the White River Group, which underlies the Ogallala throughout most of Nebraska. Beneath the Brule is the light-colored Chadron Formation, the other member of the White River Group, composed of roughly equal parts silt and volcanic ash; beneath the Chadron is black, carbon-rich Pierre shale, once the floor of the sea. Over the Brule in some areas is the Arikaree Formation, which contains even more ash than the Chadron. There is no Arikaree in Arthur County; Goeke expects to go directly from the Ogallala into the Brule about 750 feet below ground level. There the drilling will stop. The crew will pull the drill stem out of the hole and run an electric log—a graph of the electrical resistance of the well's walls, produced by slowly lowering a probe bearing a live electrode to the bottom of the well and just as slowly hauling it out again. Then they will go home.

Although everyone at the project site defers to Goeke, he is here unofficially; his current employer, the University of Nebraska Research and Extension Service, is not part of COHYST. The drill is under the nominal direction of Clint Carney, a slender young man with sunglasses, a ready smile, and a freshly minted master's degree from Northern Illinois University. Carney, his dark hair scrambled by the wind, stands at the south end of the examining table, logging the samples. He also times each drilling interval; these times depend on the resistance the bit meets, and knowing them will help determine the density of the subsurface layers. Carney is left-handed, so the watch he looks at to time the intervals is worn on his right wrist. Goeke is also wearing his watch on his right wrist, not because he is left-handed but because he is trying to protect the watch from the drilling mud that is spattering his left arm as he reaches over the pit to catch the samples. It is a futile gesture: his right arm is getting spattered, too, along with his legs, chest, face, and hard hat. After a while he takes off the watch and puts it in the glove compartment of his pickup.

The drill rig has been rented from a company called Sargent Irrigation out of Grant, Nebraska, and Chris Howard and Aaron Withington, who are operating it, are Sargent employees. Tom Downey, the manager of Sargent's small wells division, is also on the site—ostensibly to supervise Chris and Aaron, but mostly because test holes fascinate him. "Gets in your blood," he grins, when I ask him about it. "It's the uncertainty—it's always entertaining. You never know what you'll find down there." Almost every time I see him he is smiling, like a small boy who has just found out that tomorrow is Christmas and Santa Claus is real.

THE SAND HILLS ARE not only broad, they are deep; in some places, the sand extends as much as a thousand feet below the surface. It has been said that, if you scraped the dunes flat and carted away all the sand above the level of the surrounding plain, there would be enough left over to build the Sand Hills all over again. The drill moves quickly through the sand, completing each five-foot interval in a matter of a few seconds. Goeke looks annoyed. "There's no character to this drilling whatsoever," he complains. "This stuff is not consolidated; it's just loose, windblown sand, so it doesn't offer any resistance to the drill. It's boring—totally boring. We have to earn the interesting stuff."

Sand has a peculiar relationship to wind: It does not fly upon it, but bounces along the ground in front of it in low short hops, a process called *saltation*. Because of this, it tends to accumulate and move in the large, slow masses we call dunes, burying whatever is in its way: rocks, trees, clay beds, watercourses, other sands. If the climate becomes damper, the dunes will stabilize beneath vegetation, the water table will rise, and lakes and meadows will develop, pumping organic matter into the interdunal sands; if the climate dries out again, the vegetation will die and the dunes will go back to their old pattern of moving and burying. This cycle has been repeated at least three times in the twelve thousand years of the Sand Hills' existence. Goeke has drilled down through the top of a 125-foot-high dune in northeastern Grant County and found organic material—decayed plant leaves and stems—thirty feet below the base of the dune but two hundred feet above the bottom of the sand. He has drilled twenty-four feet down between two dunes in Cherry County and found the remains of a spruce forest. Nebraska's sand sea has smothered alluvial meadows, lake beds, and gravel bars, then moved on and uncovered them again. Fossilized bison tracks have been discovered, buried beneath many layers of later sands. Along their southeastern margin, the Sand Hills are bordered by Peoria loess, a fine-grained, wind-deposited soil known for its superb agricultural properties. The loess extends northwestward for several miles beneath the dunes, which have marched implacably over it.

The presence of all this sand has altered the hydrologic characteristics of the part of the Ogallala that lies beneath them. In most places, a 5 percent recharge rate would be considered extremely high; the general estimate, here in the Sand Hills, is 20 to 25 percent. "And in blowouts like this," states Clint Carney, gesturing toward a hole in the crest of the big dune to the north, "it's more than that. There have been studies that indicate it could be as high as 50 percent." Given the region's sixteen to twenty inches of annual rainfall, that would be a phenomenal eight to ten inches of recharge every year. In New Mexico and Texas, water trickles into the aquifer as slowly as 0.125 inches per year.

The unusual recharge characteristics of the Sand Hills give rise to equally unusual rivers. They are, claims Goeke, "unique streams—the most constant-flowing streams in the world." Through storm and drought, through summer and winter, through hell and high water— except that there is no high water—Sand Hills streams hold steady. They will continue to do so for as long as the sand, the rain, and the Ogallala Aquifer survive.

The Dismal River, which rises in an interdunal lake a few miles northeast of John Jensen's ranch, is a particularly striking example. The Dismal is probably the most misnamed river in America. Seen in the sunlight, it is a clear, deep, muscular stream flowing over sparkling sand, the big dunes lining its course bending gently back from it like the flanks of sleeping horses. I saw it in an uncharacteristic fog, with droplets of moisture clinging like beads of mercury to the cactuses and the yuccas and the soft white petals of the sand plums, the dunes fading upward like ghostly swans into a gray nothingness of damp atmosphere. The river was flowing bank-full. It always does. Streamflow gauges along its course register within a few percentage points of the same figure every time they are read. If streamflow gauges were capable of boredom, these would be clamoring for a different assignment.

The cause of this peculiar behavior is the river's environment. The whole course of the Dismal, source to mouth, is through the Sand Hills, and like all Sand Hills streams it is fed almost entirely by groundwater. The springs come up on its banks, boiling sand springs driven by the weight of the damp sand in the surrounding dunes, surging up natural pipes from—what, 120, 150 feet?

"Oh, the deepest I got in the Blue Pool was 113 feet," says Goeke with a dismissive wave, as if a churning spring that could swallow an eleven-story building was an everyday occurence. "These Sand Hills streams are predominantly groundwater-supported streams, and that's what makes them so interesting. Ninety-seven percent of their flow comes from groundwater. Overland runoff isn't a significant component." There is, as of yet, no serious groundwater development in the Sand Hills. The Dismal and its companion streams are what we stand to lose if development arrives.

IT IS NOW TEN O'CLOCK in the morning, the hole is 130 feet deep, and we have seen nothing but sand. Tom Downey frowns at the mud pit and goes to his pickup for a Marsh funnel, a long-spouted poly-ethelene funnel used to test the viscosity of drilling mud. The thicker the mud, the slower it will pour through the funnel. A thirty-five-second mud is considered ideal; forty, if you are drilling in loose gravel. This one is barely twenty-seven. Aaron Withington pulls two bags of dry bentonite clay from the bed of the drill rig, rips their tops open, and pours their contents into the pit. Goeke strolls up to watch.

"Here in the Sand Hills you can actually drill with straight water," he remarks. "But we'll get into gravel eventually, and it's better to be going in there with a heavier mud. Otherwise the water will just migrate out into the formation. The rule of thumb is that a hole will produce water in direct proportion to how much water it takes to drill it."

Because it makes rotary drilling practical, bentonite is one of the key ingredients that has made Ogallala development practical. A light-colored clay mineral derived from the weathering of volcanic ash, it was given its name in an 1898 paper by Wyoming state geologist Wilbur Clinton Knight, who found it in association with beds of Fort Benton shale. Like all clays, bentonite is characterized by extremely small particle size: sixty million grains of it can sit comfortably side by side on a dime. A dry gram—equivalent, roughly, to the twenty-eighth part of an ounce—has been calculated to contain a total surface area of eight million square centimeters, or about eighty-six hundred square feet. It is what happens when bentonite gets wet, though, that gives the material its usefulness. Placed in water, those tiny grains swell up, expanding to anywhere from twelve to twenty times their dry size. They also begin to stick together. If you are owned by a cat, you are probably familiar with this behavior: bentonite is the active ingredient in clumping kitty litter. It is also used as a binding ingredient in the sand-casting of foundry metals, as a pigment carrier in paint, and as an ink absorber in the recycling of used office paper or newsprint. Its most characteristic use, though, is as drilling mud, a task for which it seems almost to have been designed. Its tiny particle size makes it colloidal, meaning that it will not settle out in standing water; its swelling and clumping behaviors cause it to coat surfaces and block pores. And it is an excellent lubricant, a property that helps keep the working mechanism of a rotary drill bit operating smoothly. Long before drillers began demanding the stuff, pioneers were looking for deposits of it on their way west. They used it on the axles of their wagons to keep the wheels rolling smoothly toward Oregon.

AT 10:40 A.M. AND 155 FEET, the outlet pipe finally begins to spit out silt. Goeke notes that the drill has slowed a little. "It's the change in composition," he points out. "The bit has to chop its way through. Silt's not hard, but it provides a little resistance, because it's denser. A lot of times, when you drill these sands, you'll find silt seams in there later that you didn't even notice. You'll see them in the electric log, but you won't spot them in the samples—they're just too subtle." This one isn't subtle. The drill will be in it, more or less constantly, for the next one hundred feet.

At 265 feet, tiny pieces of gravel begin to show in the samples. By 280 feet the samples have become mostly gravel, mixed with bits of clay. We have hit the Broadwater Draw Formation, a deposit from Pliocene and early Quaternary times that overlies the Ogallala beneath much of Nebraska. The newer material lies unconformably on the old, meaning that there was a period of erosion after the Ogallala was laid down but before the Broadwater Draw arrived. I ask Goeke where the gravels came from.

"West," he says, succinctly. "This is the Rocky Mountains you're looking at."

"But this is not the aquifer?"

"This is *an* aquifer."

"It's water-bearing."

"Oh, yeah. Hell, yeah. But it's not the Ogallala."

"Okay. Do you go straight from this into the Ogallala, or is there a layer of something else in between? How can you tell them apart?"

"The Ogallala has certain indicators. There are little tubules, for example—fossilized rootlets. Small roots that have had their structure replaced by silica. You'll see a lot of them. There aren't any here." He teases a Broadwater Draw sample apart to show me. "No rootlets. Until you start seeing rootlets, I don't think you can call it the Ogallala."

"But it comes from the same place."

"Yeah. Well, the Ogallala comes from the whole Front Range, and this is just from a localized source area. There's anorthosite in here. That's a feldspar with a sort of a blue sheen to it—really distinctive, if you find some. The only place you get it is over west of Laramie, in the Snowy Range."

THE LARAMIE MOUNTAINS

Now we are headed south, following gravel roads down the floor of the flat valley that separates the High Plains from the base of the Laramies. Dust billows behind the Cherokee. Tiny towns flash by, as quick and insubstantial as text on a television news crawler.

Mason is describing a recent storm. "The radar indicated that about six inches of rain fell," he remarks. "I think the closest they came was three inches actually measured by an observer. But it caused a lot of damage, and this road was washed out in several places. Pretty neat to come out the day after and see the alluvial fans all over the place, where it had just blasted out and eroded a hill." He pulls the Cherokee off the road at the top of a small rise. "Here's a good view of the Gang-plank. We're going to get a lot closer to it, but while we're here we might as well look."

The Gangplank looks like a gangplank, rising in a smooth line from the dock of the plains to the deck of the Laramies, as if designed to lead travelers aboard. It has been used that way for centuries. Interstate 80 and the Union Pacific Railroad are only the latest routes to have taken advantage of the Gangplank's smooth surface to climb painlessly up the outermost wall of the western mountains.

"The Gangplank is probably an old stream channel," Mason continues. "You're probably looking at inverse topography. Sands and gravels are more resistant to erosion than floodplain deposits are, so streambeds eventually rise above their surroundings." I have been thinking of the Ogallala as composed of the ghosts of ancient rivers; here is such a ghost, a phantom of the past, made visible. The vision is not exact. The Laramie Range is still rising, tilting the lands that touch it; the gradient of the Gangplank is almost certainly not the same as the gradient of the stream that deposited it.

A few minutes down the road, Mason points to a gully. A flat fan of raw earth, its edges scalloped into rounded pseudopods, spreads outward from the gully's maw over the road's shoulder. "During the storm, you had headward erosion happening on all of these hills, and in all the little dry drainages coming off of them," Mason explains, "so you had water that was heavily laden with sediment. It was almost a slurry, I'm guessing, from what it looked like afterward. But as soon as it hit the floor of the valley, the water couldn't hold the material anymore, so the alluvial fan just spread out." He grins sheepishly. "You can always tell scientists, because we get excited about the littlest things. The microcosm versus the real world. Seeing what happened in this little tiny runoff event, I was thinking, 'Well, yeah, that's a scale model of how the whole High Plains must have been deposited.'"

He is silent for a while. The Cherokee angles onto Wyoming 210, purring west and uphill. A few minutes later Mason slows and turns left, onto a gravel road across the face of the Laramies. Meadows spread above and below the road; clouds of small flowers hover just above the grass, hinting at, rather than shouting, the arrival of spring. We round a corner and there is the Gangplank, top to bottom, foreshortened from this angle but unquestionably a western extension of the High Plains. Its upper end towers over us.

"Now," says Mason, "I'll tell you how the rocks got to the Holiday Inn. It has something to do with the Gangplank. I've never heard the High Plains tied to it, but I'm sure they've got to be tied."

He gestures toward the pink summits to the west. "When the Laramide Orogeny began, the mountains started to rise, and there was a lot of erosion. The debris was deposited in the intermontane basins. This went on for tens of millions of years, and by Tertiary time, all of those basins—the Bighorn Basin, the Wind River Basin, the Green River, the Laramie—all of them were completely full. All but the highest peaks were buried under erosional debris. Streams from the Snowy Range actually ran all the way across the Laramie Basin, on top of the basin fill. When you look at the Gangplank, you're not only looking at something that goes up to the Laramie Range, but a remnant of something that once went all the way across to the Snowies.

"One of the strongest pieces of evidence is the way those rivers we talked about slice right through the mountains. How did that happen? There are two possible explanations. One is that the streams were cutting the whole time the mountains were rising. That's a hard one to swallow, because it's difficult to imagine that erosion exactly kept pace with uplift. The more accepted theory is, again, that the basins were completely full of sediments. It was wind erosion, apparently, that took most of that away. As the basins deflated, the streams just cut down through the mountain ranges like a saw."

I consider this. "When the Gangplank gets to the top of the Laramie Mountains, how close is it to their summits?"

"Well," says Mason, "that goes back to the story. When you drive across on I-80, you'll notice there's a plain on top of the mountains. It's all granite, but it's pretty flat. The idea is that the granite there was planed smooth by the streams coming down from the Snowies. Then, after those streams had moved across and deposited the High Plains, there was more regional uplift. The plains uplifted, and they started to be dissected by the streams instead of being deposited. At the same time—I don't know if it was the same time. We're talking about geologic time here, so there may be a gap of a million years. But the basins deflated. The wind took the sediments out."

"Did it move them over to the plains? Where did they go?"

"Probably the Atlantic Ocean." He smiles. "In the thirties, they documented dust from Oklahoma and Texas blowing all the way to the Atlantic. That is happening today in reverse—storms in Asia can bring dust to us."

We fall silent, regarding the scene before us: the flowers and grasses, then the Gangplank, then the blue infinity of the plains, the displaced remnants of ancient mountains, full of stones and water from other places and other times. "The whole thing is about scale, you know," Mason muses quietly, after a while. "How much dust can you move in ten million years?"

THE SAND HILLS

Thirty-five feet into the Broadwater Draw gravels, the drill hits hornblende and olivine—heavy, green minerals that tend to sort toward the bottom of a gravel bed. Anticipation rises.

Two intervals later, we are back in silt. Clint Carney, logging the sample, looks frustrated. "The geology of western Nebraska is complex," he complains. "In the eastern part of the state it's pretty much layer-cake, but around North Platte it begins to get interesting. And when you get into the panhandle, it's just crazy. I'm glad I'm not assigned out there."

There is another silt interval, duly logged and bagged. The drill cuts into the subsurface for the sixty-fourth time that day. Goeke looks at the sample and smiles. It is 1:07 P.M., Mountain Daylight Time, and we are 320 feet below the surface. "Rootlets," he announces happily, pouring the sample onto the table. "And it slowed down, the last foot." He waves a hand at the small pile of gravel on the raw one-by-twelves. "Somewhere in there is the top of the Ogallala."

IT DOES NOT, FRANKLY, look like much. An amorphous mass of little damp pieces of rock, pinkish-gray in color, with flecks of maroon and dark green. Water, milky with drilling mud, oozes from it down the slight slant of the sample table. The drill has chewed whatever form it had to bits. Perhaps Goeke can reassemble the pieces in his mind and tell you what was originally gravel and what has been made gravel by the machinery, but those of us with lesser powers of observation will have to remain befuddled.

I pick up one of the rootlets. There are just a few; only the last foot of the five-foot sample interval contained them. The one I'm holding is a round white stone cylinder an eighth of an inch through and a quarter-inch long, flattened a bit at one end, as though crushed slightly before calcification. A second rootlet, just a sixteenth of an inch through but nearly an inch long, has a slight sinuosity to it, like a potato sprout. The centers of each are dark; that is where the pith has been replaced by earth, now also hardened to rock. The petrification is perfect. A close examination of the rootlets' broken ends with a hand lens reveals the regular grid of cellulose.

The top of the Ogallala Formation—it would have been the land surface at the end of Ogallala time—is remarkably uniform through-out the eight-state High Plains region. There is a slight slope to the east, about a foot per mile. There are broad, flat bulges, none very high, which may be interpreted as the lobes of stream deltas laid down in the final stages of Ogallala deposition. There are a few shallow linear depressions: these represent stream valleys, the rudiments of a surface drainage system that did not develop much before the next wave of materials, the ones making up the current surface, began to cover it.

The bottom of the formation, by contrast, is complex. Ogallala materials landed on a well-developed erosional surface. There are buried gorges and buttes. There are buried sand dunes and gravel ridges. Under Amarillo, Texas, there is an entire buried mountain range. It is usually possible to predict, within a few dozen feet, where a well will encounter the top of the Ogallala, but the assumption that today's hole will bottom out at 750 feet is just that—an assumption, not a prediction. The only way to test the assumption is to drill down and look.

So all through the afternoon the well on John Jensen's fence line creeps downward through the Ogallala Formation, one five-foot interval at a time. There are lenses of sand and clay, there is sandstone, and there is what might once have been soil. For the most part there is gravel. Hundreds of feet of gravel. The gravel always contains rootlets. The drill keeps moving downward.

Once, Goeke calls me to the sample table to show me what looks like a ping-pong ball for mice. "It's a hackberry seed," he explains. "Hackberry's a native tree. This seed is probably—what? Seven, eight, ten million years old? Hackberries are the most persistent seeds we see in the Ogallala, but there are others. There are grass seeds that come up, and they're still diaphanous, like the fresh ones. You can mistake them for grains of sand, but if you put them between your fingers, they'll pop, just like little glass balls." He holds the hackberry seed tenderly, twirling it gently between his fingertips. The seed is now stone, but it is perfect. Its sides are crenulated, like a peach pit; a raised band around its equator shows where the two halves of the shell once met.

There is a rough spot at one end. That, Goeke points out, is the place where something—a passing animal, a storm, maybe just ripening and gravity—pulled the seed away from the plant one fine summer day at least seven million years ago.

"What's the signature that tells you it's the same aquifer here as in Texas?" I ask.

"Timeline," he responds, almost absently. He is teasing apart the sample, looking for more seeds. "'Ogallala' refers to the erosional debris carried eastward from the Rocky Mountains from about twenty to three million years ago. That doesn't mean the composition will be the same everywhere, because you've got a thousand miles of different source materials. It was the same processes acting over the same time, but on a variety of rocks. The development in every area has a different signature. Up in Pierce County, we've found gravels at the base of the Ogallala, but they're not predictable. There might be forty or fifty feet of them right on the bottom, right before the Brule. See, there's another seed fragment—all kinds of rootlets—there's another seed fragment, a little bit of a good sized one." He smiles, almost reverentially. "That's what keeps me going. When this stuff gets tedious, you look for the little surprises."

BY 4:30 P.M. WE have reached a depth of 510 feet—190 feet into the Ogallala. The air has cooled and the wind has picked up. There is an occasional drop of rain. Clint Carney shrugs into a parka; Tom Downey adds a windbreaker. Only Goeke seems unaffected. Having long ago discarded the flannel shirt he was wearing at the beginning of the day, he is wandering around the site in a blue work shirt with the sleeves rolled up. Watching him, I feel a certain amount of awe. Goeke has caught every sample—over one hundred by now. He is clearly tiring; you can see that in his posture. Once he has, by his own admission, temporarily mistaken drilling mud for sample. Yet here he is, in weather that is driving much younger men under wraps, sticking his hand again and again into a four-inch stream of forty-degree ground-water to judge, by the feel of the grit riding in it, what is significant and what is not.

Weather is an occupational hazard of well drilling. When you are working in the open, boring into the naked earth, it is impossible to avoid what the earth's protective layer of air may choose to do to you. Goeke has drilled through summer days in the shadeless Sand Hills when the air temperature topped 100°F, the sand sizzled, and no breeze stirred the oppressive atmosphere. He has drilled through winter days when the frost was so solid that a backhoe couldn't break it, forcing him to pile tires on the drill site and burn them to melt the ground before he could dig the mud pit. Once, he and his crew were drilling in the middle of a flat valley in Custer County. A violent thunderstorm came up, accompanied by a Niagara of rain. They took shelter in the lee of Goeke's pickup. A lightning bolt struck so close by that the soles of their feet tingled. When the storm had passed, they found that their flat valley had become a lake, inundating the drill rig. On another occasion, they took on a multi-well project during a winter when the temperature didn't top 10°F for thirty straight days. The mud pits froze. Diesel fuel turned to jelly overnight and had to be thawed with a blowtorch before the equipment could be started. "Somewhere," Goeke has remarked mildly, "drilling situations like this gave rise to the saying about it being colder than a well digger's something or another."

Weather has also had a large role in building the Ogallala, which is the product of erosional forces, also known as "weathering." Running water wears away rock; wind with sand in it scours rock away, creating more sand and more scouring. The sun plays a major part. Objects expand when they are heated and contract when cooled, and the daily repetition of that cycle can, over time, crack stone. In the high mountains, the sun melts snow; meltwater trickles into the cracks caused by expansion and contraction. When the sun goes away at night, the water refreezes, expanding in size by roughly 9 percent. The force of that expansion, called "frost wedging," breaks the mountain apart. Pieces of it tumble downslope in the morning, when the rising sun causes the ice to release its grip. Streams and storm runoff carry the pieces away from the mountain and out onto the plain. That is how the Ogallala was made—as simple as that. It took time, but time is something the earth has plenty of.

It is not something the drilling crew has plenty of. As the clock inches around and the weather worsens, tempers grow short. Goeke looks for a clean bucket; there are no clean buckets. He is the iris in an eyeball of dirty white five-gallon buckets, and he is holding a sieveful of sample with no place to put it down.

"Hey, Clint!" he bellows angrily. "You want to rinse a damn bucket?"

Carney—nominally the boss here—hurriedly picks up a pair of buckets and sprints for the spigot on the back of the water truck. "This is what a master's degree will get you," he mutters to me as he rushes past. Goeke carries the sieve of unrinsed sample over to the table and comes to stand beside me, wiping his hands on a blue rag.

"Did you watch the behavior of the rig during that last interval?" he asks. He still sounds irritated. "It sailed through the first couple of feet, then it slowed down, and it slowed down some more during the last bit. You need to be watching. You have to become part of the damn drill rig. The best position of all is the driller's position, because when you're standing on that rig, you can actually feel the changes to within a foot or two. The person who's catching samples is the next in line, and that person has to be tied directly to the driller's hip. That person has to watch and listen and intuit everything that the drill rig is saying about the geology, so the sample can be reconstituted as it comes to the surface. Then the catcher has to take it over and give it to the fellow at the table who's going to describe it. And if the fellow describing the samples has half a clue, he's also watching the rate of penetration, and looking at where it stops and slows, and he ought to be putting marks down, and it just bothers me that Clint—" He takes a deep breath. "He doesn't seem to be engaged. And that just frustrates the hell out of me."

There is a moment's pause. The the bit moves downward.

"Hear that?" Goeke asks. "The drill will talk to you. If the material is hard, the bit will take a bite, and then it will drop down. That's that *thunk-thunk-thunk*. The *rat-a-tat-tat* was sand. Part of the problem of the Ogallala is that you get unconsolidated sands in between sandstones. You get sandstones that are consolidated, and sandstones that are poorly consolidated, and all the erosional debris. There are much coarser channels in the formation. Boulders the size of houses. That's the kind of range you get." He has been leaning on the tailgate of his pickup as he talks, rubbing his hands with the blue rag. His hands are red and swollen and chapped, and the skin on them is puckered, as if they had been in dishwater too long. Carney passes us wordlessly, a clean bucket swinging from each hand. Goeke tosses the rag into the bed of the pickup, heaves himself erect, and slouches back over to the mud pit to catch one more sample of Nebraska's underside.

Tom Downey walks by, and I stop him to ask what the crew will do if they run out of drill stem. "Quit," he says, smiling. They have brought a thousand feet along, which should be more than enough to do the job. Downey glances up at the gathering sky, from which more raindrops are beginning to fall. "You still need Mother Nature," he observes. "You can't do it all with irrigation. Mother Nature spreads it around."

AT 4:50 P.M., GOEKE approaches me. "What time is it?" he asks, slumping wearily against the door of Carney's SUV. He doesn't wait for an answer. "I am rapidly losing my enthusiasm for this project," he announces. "We might not make six hundred feet today. My hands are tired. But there's nobody else who seems capable of doing it."

Two intervals later he is back. "We hit clay," he states. "That will provide a wonderful horizon, because we were drilling through sandstone, and all of a sudden we hit a clay sill. That stuff will show up like a—like—" he gropes for an analogy and gives up. "It'll be a really good marker on the electric log. It makes a good place to stop."

Howard and Withington are pulling the drill stem, lifting it section by twenty-foot section out of the hole, unscrewing the sections from each other, and stacking them in a rack on the side of the drill rig. Goeke explains that the stem can't be left in the uncased hole overnight, because the walls might collapse and bury it. They will thread it back down in the morning and start anew. Carney is tidying up the log. Howard, who has been keeping a separate driller's log but has neglected the last few intervals, comes over to copy from Carney's. "What is this stuff?" he asks. "Dinosaur shit?" Carney grins and points to the correct name in his own notes.

"How deep did you get?" I ask Goeke.

"We got down to five-eighty," he says. "We'll finish the job tomorrow."

NORTH PLATTE

I stayed in North Platte the next day, attending to business unrelated to Arthur County. It rained heavily all day. I often wondered, as I ran my errands through the downpour, how the small cadre out on John Jensen's ranch was faring. When evening came, I called Goeke to find out.

"It was a dismal day," he stated, "but we got it done."

They had hit the bottom of the aquifer at 940 feet. "That thick!" I exclaimed.

"Yeah," he said. "A lot thicker than anybody thought. Come see me tomorrow, and I'll tell you about it."

Which he did. The main building at the University of Nebraska's

North Platte Research and Extension Center is a flat-topped, single-storied structure of glass and light-colored brick that appears to date, like many extension-service offices, from the 1960s. Goeke met me in the lobby and led me to the lone conference room, an opulent, windowless space set dead center in the building. Oiled cherrywood and teak, thick carpets, black leather high-backed chairs. The geologist was dressed as I had last seen him, in worn jeans and a flannel shirt with the tail dangling; he still looked like an unmade bed. He asked if I had enjoyed my day at the drill site.

"It was fun," I said, "watching you guys work."

Goeke grimaced. "Well, it was no fun yester—. I take that back. I condemn it, but . . ." He left the sentence unfinished. I suspect that Jim Goeke is constitutionally unable to avoid having fun while drilling, however miserable conditions may be.

"But you finished the job." I wanted to be sure.

"Oh, absolutely. Clint left about noon, thinking this was going to be a 750- or 760-foot hole, and it went another 180 feet. We hit gravel down at the bottom that no one knew was there, and a bunch of cementation that I hadn't seen before. And Clint was gone. But you don't drill a hole like that and not finish it. There's an obligation, with the investment."

We talked about the need for test holes. "You can never have too much data," insisted Goeke. "You can guess at the subsurface structure at the land surface, or you can drill a hole and try to understand what's down there." Nebraska requires drillers to keep logs on all irrigation wells, but Goeke finds these of limited use. One reason is that irrigation wells go only to the water; the aquifer may be a thousand feet thick, but if there is an adequate water supply thirty feet down, thirty feet down is where the well will stop, and the remaining 970 feet stays unknown. A second, stronger reason is the quality of the data. Drillers like Tom Downey, who really want to know what's down there, are rare. Most keep a log only because it is required of them, and their understanding of what they are drilling through is limited to how the different textures they encounter will affect the work of the drill.

"You probably know people who are drillers, and I don't want to demean them," Goeke sighed. "But they're basically mechanics who can keep a complicated piece of equipment like a drill rig working. Their job is to make a hole in the ground and pull water out of it. They don't care that they're drilling backward through time, or that what they bring up with the drill explains the history of the earth. That's what's really neat, for me—and they just hammer right through it. But that young kid, Chris, who was drilling yesterday—he saw those hackberry seeds. He's been drilling for a year or so, and he had never looked at anything close enough to see the seeds. He's going to school. It really is an educational experience when you drill into the Ogallala, because of the variability. You try to understand the sense of things, and then you get captivated."

"I have a friend back in Oregon who's a geologist," I remarked. "He says you can always tell the real geologists, because they're the people who want to go to Hell when they die."

Goeke laughed. "I don't know about that," he said. "Some of us just go to Nebraska. Or Kansas. Go to Kansas." His look turned inward. "I bought—oh, about thirty acres on the North Fork of the Dismal River. I own both sides of the valley. Old sod shanty on the south side, all fallen down. Water just all over everywhere. Everybody says, 'What are you gonna *do* with it?' I mean, it is really out in the middle of nowhere. And I say, 'I'm not gonna do *anything* with it. And neither is anybody else.'"

"Do you go up there often?" I asked him.

"Seldom. Every now and then." He leaned back and put his hands behind his head. "It just makes me feel good to own it."

XII

HOLOCENE

EOLOGY IS NOW, but it happens slowly. Plains are built
from mountains, grain by grain, over millions of years.
Rivers cut down through plains at the same infinitesimal
rate. There are occasional bursts of speed during which you can watch
geology happen: earthquakes, floods, volcanoes. These seem dramatic
to us, but in the overall scheme of things they are very small. All
human prehistory and history on the High Plains has been played on
a stage with much the same shape as the landscape we see today. Only
a few minor details have altered.

That is not the same as saying that the plains have always looked
the same throughout our presence here. Climate moves faster than
geology. When the Sand Hills were formed, they were as waterless as
the Sahara. People lived through that. People lived through the cold at
the end of the last ice age, and through the warm and damp period
immediately afterward when rivers flowed in the dry gulches of Texas
and mammoths roamed New Mexico over pond-dotted savannahs.
We lived through the disappearance of forests in Nebraska and the
establishment of prairie in their place. We lived through various
extinctions and vegetation shifts and droughts and heat and cold. His-

tory is the story of human interaction with geology and climate. Wars and kingdoms are fleeting and largely irrelevant events.

In the language of geologists, the epoch we live in is the Holocene. It follows the Pleistocene, the epoch of glaciers. Opinions differ as to when to draw the boundary, but ten thousand years B.P. (Before Present) is a widely accepted figure. Humans have been around the whole time. The Holocene is our history, and on the High Plains, our history goes very, very deep.

ALONG THE WEST SIDE of New Mexico 467, a mile north of Oasis State Park and seven miles north of the city of Portales, sprawls a defunct gravel quarry. The quarry, fifty feet deep and perhaps a quarter of a mile across, was hacked into the level surface of the Llano Estacado in the early 1930s to provide gravel for New Mexico's fledgling State Highway Department; it was last worked in 1978. The small A-frame structure on the quarry's south bank dates from 1993. It is built of plywood and corrugated plastic roofing, and it looks hastily thrown together, as if erected in the face of a rapidly approaching storm. Inside, a few inches from Joanne Dickenson's sneaker-clad right foot, gapes a hole in the ground. It is the upper end of a hand-dug well, and it is perhaps thirteen thousand years old.

Joanne Dickenson is a solidly built, grandmotherly looking woman with mild blue eyes, a shock of silver curls, and a master's degree in archaeology from Eastern New Mexico University in Portales. The ancient well belongs to the university. It was purchased in 1979, along with the remainder of the site known as Blackwater Locality #1. Fifty years before that, a fourteen-year-old boy named Ridgely Whiteman had stumbled across a stone spear point in Blackwater Draw, the shallow depression in the Llano Estacado that would later become the site of the quarry. An argument with a friend over the age of his find prompted Whiteman to box up the spear point and a fragment of a large bone he had found nearby and send them off to the Smithsonian Institution in Washington, D.C. Nothing much happened. Fortunately, Whiteman had also shown the materials to a Clovis businessman, A. W. Anderson, who went out to Blackwater Draw and picked up a few bones and spear points himself. Anderson's bones and spear points eventually found their way to Dr. Edgar B. Howard, a visiting archaeology professor from the University of Pennsylvania, whose preliminary examination of the site led him to believe that the businessman and the kid were on to something. He scheduled excavations for the next field season.

That was in August 1932. In September 1932 the highway department began quarrying, using a bucket-shaped Fresno scraper drawn by two horses. Almost immediately, huge bones began to turn up. These were put on display in the window of Ed Neer's store on Main Street in Portales (ED J. NEER. FURNITURE—UNDERTAKER. DRUG'S.), where they attracted crowds of the curious. Portales civic leaders began to wonder if they could turn Blackwater Draw into a tourist attraction.

In the meantime, though, the work would go on; and so it was that, when he returned in the spring of 1933 to begin his archaeological dig, Ed Howard found himself in a race with a dig of another kind. The race continued for forty-five years, with Howard and his successors pitting themselves first against horse-powered equipment and men with shovels, later against bulldozers and backhoes. It was a contest the archaeologists felt they had to win. The urgency came from the site's stratigraphy, which had established, very rapidly and beyond any reasonable doubt, that the delicately fluted "Clovis points" found in the draw had been created and used at the close of the Pleistocene. That made them the oldest datable human artifacts in all of North America.

The quarry operators—as enthralled with the Blackwater Draw finds as most of the rest of America—tried hard to meet the archaeologist's needs. They changed their schedules to accommodate the digs, lent their machinery to clear off overburden, diverted springs uncov-

ered by the quarrying so the spring runoff would not inundate important cultural materials, and generally supported the work as much as
they felt able. However, as New Mexico writer Lienke Katz has pointed
out, they "also had a business to run, in which time equates to money."
So when one of the heavy-equipment operators found what he
described to his wife as "something really pretty . . . a perfectly round
hole, with layers of colored sand," he quickly fired up his bulldozer and
covered the thing over with dirt, afraid that if he reported it, quarrying
at the site would stop once more. Which it did, when archaeologist
Shirley East found the same hole—or a similar one—a short time later,
in 1964. Identified as an ancient well and dated to Clovis time, the
"really pretty" hole helped convince the New Mexico legislature to protect the site. The money came through in 1978. Within a year, Eastern
New Mexico University had purchased the quarry and closed it down.

In the meantime, though, the well had been lost. Not destroyed, just
lost. East and her colleagues had measured it and then backfilled it for
protection, and now no one could remember precisely where it was.

Determined work in 1993 by Dickenson—by then the site's curator—and Vance Haynes, an archaeologist from the University of Arizona, solved that problem. Using maps and data from East's 1964
report, Dickenson and Haynes re-laid her grids, re-uncovered the
well, and re-excavated it. Dave Meltzer, a wiry little anthropologist
from Southern Methodist University, got down in the hole and did the
actual digging. "He was the right size," smiles Dickenson, "and he was
visiting the site and volunteered his services." The A-frame was erected
shortly afterward. Protection, this time, was not going to lead to
disappearance.

How do they know it was a well? "It was first mapped as an
'unknown unconformity.'" Dickenson explains, her curls brushing the
translucent roof of the little protective building. "But we've found stri-
ations, identifiable as tool marks, where they cut through the caliche.
You can see that the hole's a lot smaller there. When they got into that

stuff, they didn't cut any bigger than they had to." She indicates a narrow throat three feet below ground level where the well pinches in to pass through a foot-thick layer of hard white calcification. Below the mineral deposit the hole bells out again, penetrating roughly eighteen inches into the top of the Ogallala Formation. Off to one side, at what would have been ground level during Clovis time, Dickenson and Haynes found a pile of dirt, identifiable as the spoil from the well. Four turtle shells lay nearby. It is impossible to avoid speculating that these are some of the well-diggers' tools, set down after use and not picked up again for more than thirteen millennia.

How certain is that thirteen-millennia age? Dickenson shrugs. "How do you date a hole?" she asks rhetorically. "But there was a fifty-year drought that has been reliably dated to around thirteen thousand years ago. The pond dried up. The well was probably dug at that time. A lot of mammoths died then—they couldn't dig wells."

The hole at Dickenson's feet is not the only well that has been found at Blackwater Locality #1. Eighteen others have since turned up, most of them in the crucial years between 1964, when the first well was found, and the closing of the quarry. Nearly all were subsequently destroyed. At least two showed evidence of use: They had been carefully lined with red clay. Beyond that, little can be said about them. They were found at sunset, directly in the path of the bulldozers. By the time the archaeologists got back to the site the next morning, the clay-lined wells were gone.

And Ridgely Whiteman? In 1997, Lienke Katz managed to track him down: a spry eighty-three, he was living quietly near Clovis, only a few miles from the site of his famous 1929 find. Whiteman stated that he had helped Ed Howard with the original Clovis point excavations in 1933. "Dr. Howard was a gentleman of an introverted disposition," the old man recalled. "I could never make him laugh, or even smile." Or give proper credit to the site's discoverer. Whiteman's role wasn't acknowledged until many years after Howard's work, when the

Smithsonian ran across the materials a boy had mailed to them in 1929 and belatedly recognized them for what they were.

BLACKWATER LOCALITY #1's wells are not the only prehistoric wells dug into the Ogallala, just the oldest. They may not be even that. There are a number of other ancient wells—"paleowells," as archaeologists call them—scattered across the High Plains. The ages of most of these cannot be determined. As Dickenson says, how do you date a hole? People dig wells for water, not for the convenience of later archaeologists. It is relatively easy to date an aquifer. It is much more difficult to date a space where part of the aquifer has been removed to get water out of it.

The diggers of these wells remain mysterious, but a few hazy details have begun to emerge. The first human users of the Ogallala Aquifer were hunters and gatherers who probably established seasonal camps near the migration routes of large herd animals, notably mammoths and *Bison antiquis* (a bigger, hairier version of today's buffalo). They appear to have kept their residences and hunting camps separate, butchering the meat before taking it back to their villages: Enormous piles of bones, along with scattered knives, hide scrapers, spear-shaft wrenches, and other hunters' tools have been found at Blackwater Draw, but there is no evidence of hearths or lodges. The animals were brought down by spears, possibly launched by *atlatls*. The points were not fastened directly to the spear shafts, but to short foreshafts of wood or bone that were bound to the spears: A good spear shaft was hard to come by, but foreshafts could be made quickly and easily. Clovis hunters probably carried several foreshafts with points already mounted to them. If one broke off in an animal's body, they could quickly tie on another and keep working.

There is little evidence for the use of plant foods, but that may be because we know only these people's hunting camps, not their residence areas. Domestic items are also lacking, probably for the same reason. Clothing would have been made primarily from animal skins, but it was not crude. A bone sewing needle found in 1968 in a Clovis-era rock shelter in the state of Washington was the same size and shape as its modern counterparts, with a tiny, perfect eye; any sewer today could pick it up and use it without a second's hesitation.

Perhaps the sweetest window into how the Clovis people actually lived is provided by a pair of flint knives, one large and one very small, found together near a mammoth vertebra at one of Blackwater Locality #1's meat-processing areas. Where did the little knife come from? Joanne Dickenson smiles. "They used the points down to nubs and then reworked them," she notes, "which might explain the size. But I think, with that little one, it was probably a kid practicing."

A delicious thought. One pictures a grown-up and a small child working side by side, the adult cutting meat efficiently away from the vertebrae, the child watching and copying, or taking instruction, or maybe simply sticking the toy point into the animal's body over and over just to see it slide in and out. Someone calls them; they lay their tools down side by side and walk away. And never come back.

CLOVIS POINTS APPEAR to have been an invention of the southern High Plains, born of the interaction between humans, large animals, and the huge, flat land above the Ogallala Aquifer. They spread rapidly through the hemisphere—samples have been found from northern South America to above the Arctic Circle—but their heaviest use was in the plains environment that gave them birth. If you map known Clovis sites, there will be a light peppering of dots throughout the continental United States, in all Canadian provinces, and all the way down into Colombia and Venezuela. There will be a thick smudge of dots on the western Great Plains. Draw a line around this smudge, and you will have a fair representation of the boundary of the Ogallala Aquifer. This is probably not a coincidence. The great herds the Clovis hunters followed needed both open space and water. The land over the Ogallala, then as now, would have met those twin needs better than any other place in the Western Hemisphere.

Clovis culture appears to have lasted, in its pure form, for only a few hundred years. The characteristic Clovis point, made with a flute—a shallow groove—to ease attachment to a shaft, was rapidly replaced by regionally evolved variants. Remarkably, one of the first features lost in this evolution was the flute itself. We can only speculate why. Perhaps the invention of the bow and arrow had something to do with it. Perhaps the flutes eased the penetration of the points through the hides of large animals and were no longer necessary when the largest of these—the mammoth and the antique bison—disappeared. What is certain is that the flutes lasted longer in the land of their birth, the High Plains, than anywhere else. They were gone from most other regions by ten thousand years B.P. On this long land with its great herds and ghosts of ancient rains, they remained in use for another two millennia.

WHEN CLOVIS CULTURE broke down, it was succeeded by a swirl of regional styles that persisted, in shifting form, for thousands of years. By the dawn of the modern era, though—the era of European contact—a unified Plains culture had once again emerged. Details varied through time and among different groups, but the general outline was always the same. It included two distinctive features. One was dependence on the buffalo. The other was dependence on the Ogallala Aquifer.

Dependence on the buffalo is cited in every ethnographic study of the Plains peoples. The animal has been referred to as the "Plains supermarket." Its meat provided food, its hair and sinews provided thread and twine, its hide provided clothing and home-building materials. Its bones made tools to process its own body into meat and sinews and hair and hide. Buffalo skins were tanned in buffalo urine; buffalo meat was cooked over buffalo-chip fires. Buffalo tallow was used to waterproof buffalo-skin houses. Plains babies were conceived on buffalo robes and swaddled in them following birth. There is no mystery why the buffalo figured so prominently in Plains religious rituals; the mystery would be if it did not.

Dependence on the Ogallala is less obvious. Water is water, necessary but prosaic: You can use it wherever you find it, provided it is clean enough. That statement is undeniably true, but it is also true that, almost everywhere you find water on the High Plains, it has been put there by the Ogallala Aquifer.

Rivers are no exception to this rule. The vast majority of plains rivers originate on the plains themselves, born in springs bubbling up from the Ogallala's deep reservoir of ancient rains. There are just three streams that course all the way across the plains from the Rockies—the Platte, the Arkansas, and the Canadian—and they count on the Ogallala to keep them full on the long, dry traverse. Except in times of flood, only a small portion of the water that leaves a plains river's headwaters reaches its mouth. The rest either evaporates or sinks into the ground to provide local recharge, reemerging as baseflow after months, years, decades, or centuries. Aquifers are the governing mechanism of the hydrologic cycle; they smooth out the movement of water, assuring that some will always be present, even between rains. And the only significant aquifer, the only effective hydrologic governor throughout the entire eight-state High Plains region, is the Ogallala.

It is wrong to state that nature and humans were in balance on the plains when Europeans arrived. Nature is never actually in balance; there is a dynamic give-and-take among natural processes that keeps them always off-balance, always in motion, and always evolving. What *is* true is that there were large negative-feedback loops operating on the plains that damped out major fluctuations in the environment and gave nature, and the human cultures that depended upon it, a certain sense of long-term predictability. What was present one year could be counted on, within limits, to be present the next, and the next, and the next. Change came slowly enough to manage.

Earthquakes of buffalo—the last of the great herd animals of this American Serengeti—rumbled between water sources, cropping the short grass. Hunters followed them. Because they were followed, the buffalo moved on; because the buffalo moved on, streambanks and grasses recovered quickly; because streambanks and grasses recovered quickly, the scant rainfall infiltrated the soil and recharged the aquifer instead of running into the streams and away from the plains, and both the buffalo and the hunters always had grass and water. The aquifer was in dynamic equilibrium, with about as much flowing in as flowed out. It was this shifting, stable relationship of herd, hunter, grass, and water that the arrival of European culture would permanently disrupt.

THE FIRST DISRUPTION WAS the introduction of the horse. Clovis people had known horses, but they had eaten them; hunting pressure probably contributed to the horse's extinction in North America near the close of the Pleistocene, along with the mammoth, the mastodon, the ancient bison, and the American camel. In 2001, University of Calgary scientists found the butchered remains of one of the last prehistoric American horses beneath the floor of a reservoir in western Alberta, along with the spear points that had killed it. There was no question about the connection between the bones and the weapons: The points, which were Clovis style, had protein residues from the horse's meat clinging to them.

Horses didn't return to the High Plains until the Spaniards brought them in the mid-sixteenth century. The Plains peoples quickly grasped the opportunities the animals offered, and stole them. A new culture emerged, one that bore a striking resemblance to that of the Mongols, the horse nomads of the Asian steppes. There was a revolution in hunting practices. Buffalo hunts had been communal affairs, with beaters driving the animals into enclosures where the hunters could easily pick the animals off. The horse offered a single hunter the ability to chase down a fleeing animal and kill it by himself, a task that was made even easier if he could lay his hands on a rifle. Much of the ritual surrounding the hunt suddenly became obsolete. The relationship between bison and human began to fray.

By the end of the nineteenth century, it had unraveled completely. The horse began the job; the rifle continued it; the railroad finished it. Railroads cut the buffalo range into segments too small to support the great herds. They also offered a rapid means of getting buffalo meat and skins to eastern markets, encouraging the growth of a buffalo-hunting industry. The *coup de grâce* came when the U.S. Army decided that the proper way to deal with what it termed the "Indian problem" was to eliminate the Native Americans' food supply, forcing them to choose between starvation and reservation life. Army sharpshooters began to systematically slaughter every buffalo they could find. In the 1840s, the animals had numbered in the millions; there were credible reports of herds that took an entire day to pass a single observer. Sixty years later, there were less than a thousand buffalo left.

An ecosystem is not only the sum of its parts, it is also the sum of the relationships among those parts. Cattle soon replaced buffalo as consumers of plains grasses, but cattle have a different relationship to water than buffalo do. Buffalo, wary of predators, come to water holes only to drink; cattle, the wariness bred out of them, hang around water holes and trample the life out of them. Buffalo cool themselves in the damp earth of wallows; cattle cool themselves with their drinking water. When cattle replaced buffalo, the earth surrounding the springs became compacted and rivers became dirtier. The springs stopped flowing; pores in the riverbed gravels plugged with silt, reducing aquifer recharge. These were minor changes, but minor changes add up. As farming began over the Ogallala, the underground ocean's relationship to the surface above had already begun to deteriorate. It was about to deteriorate even more.

XIII

"WHOLLY UNFIT FOR CULTIVATION"

THE DRIVE FROM KANSAS CITY to Denver along Interstate 70 takes nine hours, during nearly all of which you will be out of sight of land. The ocean of plains billows out indefinitely to either side, level and open, an immense cultivated emptiness that has little in common with conventional ideas of geography. With few other features to divert you, the freeway becomes your reality. Towns swing by as white names on green signs, as phantasmagoric as the land and as transitory as the clouds that hang on the horizon to tantalize you with dreams of distant mountains.

The town of Colby, Kansas, lies roughly midway along this journey: five hours from Kansas City, four from Denver. If you get off at the main Colby exit to buy gas, or to eat or sleep at one of the look-alike chain establishments that flock around the off-ramp, you might think you have encountered nothing more than the flotsam of the prairie sea. Venture north along Range Avenue, though, and you will enter a different world. Colby is quintessential small-town America, as real as the bunchgrass sod that J. R. Colby broke back in 1885 when he founded the place. It is perhaps the perfect prairie town. Trim residential streets step neatly back from clean brick downtown blocks that look unchanged since 1925. There are green parks—one with a small

lake—and there is a lazy, thicket-lined little river called Prairie Dog
Creek. There is a community college with an inviting campus and an
alumnus, Daniel Cormier, on the 2004 U.S. Olympic wrestling team.
There is a small symphony orchestra. And back by the freeway, visible
if you pull your eyes off the tailpipe of the car ahead, there is one of
the genuine treasures of Interstate 70: the Prairie Museum of Art and
History.

I toured the Prairie Museum's striking earth-bermed concrete-and-
glass exhibit hall on a dark, gloomy spring Saturday afternoon that
threatened a rare High Plains rainstorm. There were exhibits of
ceramics and glassware and bridal gowns, and a collection of historic
toys, many of which I remembered playing with when they were new,
dammit, and a whimsical gathering of twenty-six glass cases, each case
holding items beginning with a different letter of the alphabet (X was
a bit of a stretch). There was a room devoted to the work of opera
basso Samuel Ramey, who grew up in Colby. Outside, I wandered
among the historic buildings on the museum's extensive grounds: a
schoolhouse, a church, two barns, residences, outbuildings. The
threatened rain turned real, pouring from the black sky in gushers and
torrents. With no ark in sight, I took refuge in what might seem an
unlikely shelter: a reconstructed sod house.

THERE IS AN OLD NEBRASKA song that begins this way:

Nebraska land, Nebraska land,
As on thy desert soil I stand
And look away across the plains,
I wonder why it never rains.

When they weren't wondering why it never rained, early High
Plains settlers probably were wondering why they had come there at
all. Uncle Sam was giving away land—the Homestead Act, signed into

law by President Abraham Lincoln in 1862, offered 160-acre parcels free to any adult who could survive on one for five years—but there were serious doubts that the land was worth the trouble. The plains were horribly hot in summer and desperately cold in winter. A few trees grew along the watercourses, but there were none at all on the open prairie. The wind blew almost constantly. Rivers were trickles of water over broad patches of dry gravel, unless the rains came, in which case they turned into ephemeral Amazons. Fortunately, the rains were few. Unfortunately, the rains were too few. Agriculture was marginal. Because the soil was superb, a damp year would bring a bumper crop, but there were not many damp years, and the years between them were misery. A story is told of a traveler who visited a High Plains farm and was served corn on the cob for lunch. When the traveler asked how the crop had been that year, the farmer leaned back in his chair, picked his teeth, and remarked, "We just ate four acres of it."

Lacking wood for fuel, the settlers burned just about anything else they could find: sunflower stems, corn stalks, corncobs, grass. When crop prices were low, they burned their crops. Pieces of dried buffalo dung ("buffalo chips" or "prairie coal") burned well and were widely sought. After the buffalo were gone, cow chips—which burn almost as well—were used instead. In the days of the great cattle drives, homesteaders and cattle drovers were often at odds with each other, but that didn't stop homesteaders from encouraging drovers to bed their herds down nearby. The payoff was next winter's fuel, deposited in thick, steaming piles all over the south forty.

The homes themselves—by law, at least twelve feet wide by sixteen feet long—were usually built of sod, like my refuge from the rain at the Prairie Museum of Art and History. Living inside what is essentially a large pile of dirt may not seem entirely sane, but for the prairie, it made perfect sense. As Nebraska folklorist Roger Welsch has written, it was ". . . not only an adequate solution but also a comfortable one. It transcended the level of mere shelter and became a home."

Sod houses enjoy a questionable reputation that is only questionably deserved. A sod house (or "soddy") was extremely well insulated—warm in winter and cool in summer. It could be built quickly, anywhere, with a minimum of tools. Tornadoes rarely damaged the sod walls, which did a better job of protecting their occupants than the flimsy frames of the wooden houses that eventually replaced them. And the soddy's reputation for being small, dark, and squalid was ill-deserved. These characteristics were less a result of the building materials than they were of the financial state of the builders. Windows could be framed into soddies as easily as they could into wood homes, but windows cost money. Roof beams also cost money; shorter ones limited house size but kept costs down. When built to a large, comfortable scale, a sod house was, well, a large, comfortable house. There were L-shaped soddies, rambling soddies with long, shaded porches, and two-story soddies that looked like the dignified brick manses of eastern cities. Sod walls could be plastered and wallpapered, if you could afford plaster and wallpaper. Even the notoriously leaky sod roof was more a result of poor construction than it was of poor material. Sod on the roof actually had two advantages over shingles: It completed the insulation of the interior, and it gave the roof heft so the plains winds couldn't lift it off. Laid over a wooden subroof that was tight, well-braced, and lined with overlapping runs of tar paper—as those who could afford to usually did—sod roof tiles were no more likely to leak than were shingles. Laid on bare slats, though, they did leak—mightily. Accounts of sod-house life often mention sheets of canvas strung up over the beds, the cookstove, and the dining-room table, keeping these important areas dry while the rest of the house suffered drips—and sometimes small running streams—of muddy rainwater.

Sod bricks were cut twice as long as they were wide, usually one foot by two feet, and were four to six inches thick. The cutting was done by a grasshopper plow, a small, straight-bladed plow made without a moldboard so it would slice the earth but not turn it over. As the plow bit into the soil, there was a sound that has been described as "like a zipper opening." It was the death wail of a ten-thousand-year-old ecosystem, the sound of corded roots in the thick bundled turf of the prairie being severed. Throughout the closing years of the nineteenth century and the opening years of the twentieth that sound spread widely, until it had been heard nearly everywhere between the Mississippi River and the Rocky Mountains.

EARLIER IN THE NINETEENTH century, this region had been the Great American Desert. Travelers hastened through it; settlers shunned it. There were serious proposals to turn all of it into a permanent refuge for "savage nomads." Edwin James, the chronicler of an 1819 U.S. Army survey expedition led by Major Stephen H. Long, captured the prevailing opinion perfectly in what has become a widely quoted passage (often attributed, wrongly, to Long):

> In regard to this extensive section of country, I do not hesitate in giving the opinion, that it is almost wholly unfit for cultivation, and of course uninhabitable by a people depending upon agriculture for their subsistence.

The transcontinental railroads changed that. The change was deliberate; the various railroad companies that were vying to knit the nation's Pacific and Atlantic states together all actively encouraged High Plains settlement. It was simply good business. The railroads had customers in the East and customers in the West, but there was a big, empty customer void in the middle. The railroads would be more profitable if there were customers there, too. And "savage nomads" didn't ride the train.

So plains settlement began, driven by corporate promotion and justified by a quasi-scientific theory: Rain follows the plow. Break enough prairie, and the climate will change. Never mind that there was not enough water: The advancing tide of civilization would bring the water with it.

The slogan generated its own odd logic, because once in place, it had to be justified. Turning over the soil, it was widely believed, would keep rainfall from running off into the rivers, leaving more for plants to transpire, which would put the water back into the air where it could fall again as rain. Planting trees would do the job even better, because a tree's high ratio of leaf surface to stem diameter maximizes the interface between its respiratory system and the surrounding atmosphere and allows the transpiration of more water per unit area of ground. Even railroad building might help. "It is noted as a singular fact," gushed a reporter for the *Omaha Daily Bee* in 1886, "that the building and operation of railroad and telegraph lines, even in advance of settlement, is generally followed by a steadily increasing rainfall." The newspaper attributed this affect to the leakage of electricity from telegraph wires and to "the disturbance of the atmosphere intendant upon the rushing of the trains."

It was a strange set of notions, but it was endorsed by some influential people, including geologist C. W. Wilber (who coined "rain follows the plow") and professor Samuel Aughey, the head of the natural sciences department (actually, he was the entire natural sciences department) at the new University of Nebraska. Nature even went along with the gag for a while: Rainfall on the Great Plains increased steadily during the 1870s and early 1880s. Then the bottom fell out. The year following the *Daily Bee*'s confident prediction of wetter years ahead, 1887, turned out to be the driest year in Nebraska's history up to that time. It was followed by the Dry Nineties. During that harrowing decade, crops shrunk to less than a bushel per acre and 184,000 homesteaders abandoned their claims in Kansas alone. The talk of beneficial climate change was effectively quashed. Something else would be needed to encourage settlement on the High Plains. That "something else" was irrigation.

XIV

FINDING THE WATER

D EEP WITHIN KANSAS'S Ladder Creek Canyon, between
the placid waters of Lake Scott and the little beetles of Big
Springs, sprawls a hidden prairie nearly a mile across. Fertile, protected from winds, and blessed by numerous Ogallala springs,
it must have seemed ideal to Herbert and Eliza Steele when they filed
for a homestead there in 1888. The Steeles watered their crops from
the canyon's springs, but they were not the first. Sweating their ditches
through dense thickets of buffalo currant and hackberry, they found
themselves following the courses of ancient, overgrown irrigation
works that someone unknown had dug long, long before.

In 1898, in a field near their home, they found the answer to the
mystery: the foundation of a southwestern-style Native American
pueblo. Excavations by University of Kansas archaeologists, plus a
search of historic records, quickly identified the ruin. It was called El
Cuartelejo, and it had been built in the early seventeenth century by
Taos Indians fleeing the Spanish conquest of New Mexico. Here on
this sheltered prairie they had built a small village, planted corn and
beans, and dug ditches to their fields from the springs at the base of
the bluffs—the northernmost of the pueblos, and the first known use
of the Ogallala Aquifer for irrigation.

Herb and Eliza Steele's use of the same springs may have been the second. Groundwater irrigation was slow to take hold on the High Plains. There were a number of reasons for this, but two stand out. The first was a lingering belief that rain would follow the plow; to those who felt that way, irrigation was a temporary, strictly limited expedient, and the less it was used, the better. The second was the occurrence, here and there on the High Plains, of what certainly looked like big rivers that could be readily developed as water sources. If you were planning to irrigate, river water seemed the logical thing to use. What the settlers had so far failed to grasp was that the river water was no longer there.

Actually, there wasn't much there even in the beginning. High Plains rivers are deceptive; the water flows over their sands in thin sheets, visually broad but volumetrically challenged. It was famously said of the Platte that its waters were "a mile wide and an inch deep," a statement that is only a small stretch of the facts: The historical record indicates that, near its mouth and in its natural state, the Platte was actually a half-mile wide and two feet deep. The river's name means "flat" in French. The Omaha people, who lived on its lower reaches, called it *Nebraska*, "flat water," a name that was eventually applied, appropriately, to the whole region. When Ramsey Crooks, Robert Stuart, and their bedraggled little band of Astor Company fur traders came upon the North Platte in December 1812 in what is now eastern Wyoming, they spent more than two months building canoes in which they planned to float comfortably down to the Missouri. They launched on March 8, 1813. Barely three miles and less than a day downstream—nearly all of which they had spent dragging heavy canoes over sand shoals and gravel bars in ankle-deep water—they abandoned the boats and started walking again.

South of the Platte is the Arkansas. In the nineteenth century, the Arkansas was classed as navigable all the way to Kinsley, Kansas, less than forty miles downstream from Dodge City. This may charitably be described as an exaggeration. Here is how Francis Parkman found the river in the late summer of 1846, jaunting through on the post-Harvard lark whose journal would later become the classic travel book *The Oregon Trail*:

> The Arkansas at this point, and for several hundred miles below, is nothing but a broad sand-bed, over which glide a few scanty threads of water, now and then expanding into wide shallows. At several places, during the autumn, the water sinks into the sand and disappears altogether.

Facts rarely stop promoters, of course, nor do they blunt hope. Thirty years after Parkman's visit, not far from the location he had described, a group of energetic pioneers laid out a town they called Garden City. To make their Garden bloom, they built irrigation canals: the Garden City Canal, the Southwestern Canal, the South Side Canal, the Farmer's Ditch. The biggest, the Amazon Canal, was more than one hundred miles long. The river at that time was said to "not vary more than five feet between high and low water," which was pretty impressive if you didn't know that the river was rarely more than five feet deep anywhere along its course. Stories hailing Garden City's promise as the "center and inspiration of irrigation development" began appearing in the eastern press. Some of the stories were true.

The high point of the canal era in southwest Kansas may reasonably be said to have occurred in 1886. That was when patent-medicine mogul Asa Soule arrived, looking for a profitable place to spend several million dollars. Soule had read the Garden City puff pieces and had apparently believed all of them. Founding a new town on the north bank of the Arkansas twenty-five miles below Garden City, he named the place after Kansas's boosterish senator, John J. Ingalls (nothing like currying a little favor), and started building a ditch of his own. The Eureka Canal eventually ran ninety-six miles, from

Ingalls northeastward through two downstream counties to the putative head of Arkansas navigation at Kinsley. You can still trace much of its course today.

Having a town and a canal wasn't enough for Soule, though: He wanted a whole county. With bribes, promises, outright vote purchases, and—eventually—a ballot-box heist carried off by a posse of hired guns that may have included legendary frontier gunfighter Bat Masterson (myth-busting historians like to point out that the Masterson involved was probably Bat's younger brother, Jim), Soule attempted to snatch the seat of Gray County from nearby Cimarron. When that failed, he pulled out, abruptly, selling his ditch to a group of local investors. They never made a dime. By then, the Arkansas had essentially stopped flowing into Kansas at all; irrigators in Colorado were siphoning off the entire river. They were doing the same to the South Platte, which was "flowing" into Nebraska bone-dry. With the notable exception of the North Platte, rivers would henceforth provide no significant irrigation whatsoever on the High Plains.

That left the underground ocean. Full-scale exploitation of the Ogallala Aquifer was about to begin.

EDDIE JOE GUFFEE IS A LEAN, leathery sixty-five-year-old who looks a bit like Garrison Keillor, if Keillor had just spent six months driving cattle up the old Chisholm Trail. When I met Guffee, he was carrying a couple of coffees and a small box of doughnuts toward the entrance of the Museum of the Llano Estacado, across the street from the campus of Wayland Baptist University in Plainview, Texas. The museum is Guffee's baby: It is an offshoot of the university, where he teaches history and earth science, and he has been its director since its doors first opened in 1976. That was the same year Guffee came home to Plainview, where he had grown up, with a brand-new master's degree from West Texas State University in Canyon.

"You want to drive west on US 70," he told me. "A couple of miles beyond the interstate you'll drop down into a sort of a little valley. That's Running Water Draw. Just before you go up the far side, there's a road off to the right. Look for the first house on the right-hand side along that road. The first well's right there in the front yard." By "first well," Guffee really meant the first well—the earliest known irrigation well on the Llano Estacado.

I drove out to Running Water Draw later that day. The first house on the right was a modest brick ranchette with a hip roof and a large Ford pickup parked in the driveway. The well was right there in the front yard, under a crumbling concrete cap. A few years ago, the site was surmounted by a somewhat incongruous wishing well. "They found out it was the first well and put that up to mark it," Guffee chuckled. "I don't think they had any idea what kind of a well it was. Thought about buckets and hand cranks and that kind of stuff." Today it is topped by a small decorative windmill and a large American flag. A few yards away is the edge of an irrigated field, its old-fashioned ditch-and-furrow setup redeemed by gated pipe.

The Plainview well was not the first irrigation well drilled into the Ogallala. That honor apparently goes to a hole sunk some time early in 1910 in New Mexico's Portales Valley, very close—though no one knew it at the time—to the Clovis-era well in Blackwater Draw. The Portales Valley operation, however, was a financial failure that closed down after only a few years. The Plainview well was the real beginning. The era of groundwater irrigation from the Ogallala Aquifer may fairly be said to have started right here in Running Water Draw on January 6, 1911.

When the well came in, Plainview had been a town for less than a quarter of a century. It had first been called Runningwater, after the draw. There wasn't a great deal of water in it even then (there is none today), but the town's boosters felt the name might dispel the popular notion of the Texas Panhandle as a godforsaken desert where no one could possibly want to live. When honesty compelled abandonment of that idea, someone suggested Hackberry Grove, after the few scraggly hackberry trees growing on a low knoll near what would eventually become the center of town. Honesty still shook its head disapprovingly: There were no other hackberries for miles in any direction, and the little copse on the knoll was hardly big enough to be called a grove. So the town became, for unarguable reasons, Plainview. In 1888—the same year Herb and Eliza Steele first stumbled across those mysterious irrigation ditches in the canyon of Ladder Creek—Plainview became the county seat of Hale County.

By 1910, the growing community had three thousand people, ninety businesses, two colleges, numerous churches, a boosterish nickname—The Athens of West Texas—and a branch of the Santa Fe Railroad with an attractive brick station next to a big playa at the north end of Broadway. It also had a problem. The great panhandle cattle ranches of the late nineteenth century, on which the town's economy had been founded, were fading fast; farming was the coming thing. The soils around Plainview were rich, but they lacked water. Rainfall wasn't going to provide it, and neither was Running Water Draw. The only hope was underground.

In the summer of 1910, a group of Plainview businessmen who had heard about the well in the Portales Valley traveled to New Mexico to take a look. What they saw convinced them that an irrigation well would be worth trying in Texas, so a committee went calling on John Henry Slaton, the president of the First National Bank of Plainview. Slaton was a classic self-made man, an ex-Louisiana boy who had come to the panhandle at the age of fifteen, taken a job as a cowhand, and proceeded to work his way up. He agreed to put up the money for a test well, but with one catch: It had to be drilled on Slaton's own farm, five miles west of Plainview. If the well came in, the banker would write off the loan. If it was a dry hole, the committee would have to pay him back.

The committee looked over Slaton's farm and chose the site in Running Water Draw, twenty feet below the general level of the Llano Estacado. That would give them a twenty-foot head start. Work began in mid-December 1910 and proceeded downward through three stages to the 130-foot level, where a perforated casing and a pump were installed. The first measured flow was only two hundred gallons per minute, but the well was just clearing its throat: After it had coughed up a couple freight cars' worth of sand, it peaked out at fifteen hundred gallons per minute. Plainview residents hadn't seen that much water in a long, long time.

Slaton and the well's promoters formed a corporation, the Texas Land & Development Company. Then they approached the Santa Fe Railroad. The railroad was happy to help; trains were still trying to build a customer base in a region where there were not only not very many customers but not very many people. The deal that was worked out offered low-cost, or even free, excursion-coach tickets from Mason City, Iowa, to Plainview. That got the rubes to town. The next trick was to keep them there.

THE MUSEUM OF the Llano Estacado has a strict no-smoking policy, which Eddie Guffee approved but cannot fully follow. As we spoke in the museum's lobby, he stepped to the glass entrance door, opened it a crack, and lit a cigarette, extending his arm through the door to keep the smoke outside between puffs. "After you look at the well," he told me, "make sure you drive by the depot and take a look at Lake Plainview. There's no lake there now, but you can see where it was." Lake Plainview—the big, dry playa behind the depot—was one of the Texas Land & Development Company's principal ploys to keep the Iowans from going back home to Mason City. "Folks'd get off the train," Guffee explained, "and there was a big lake right across from the depot—sailboats going back and forth, people swimming and fishing, that kind of thing. It was real attractive." It was also a fraud. Beside the depot, to the right as you face the building from the playa, crouched a hidden well. A steam pump and Ogallala water were keeping Lake Plainview full. Guffee smiled. "They'd get off the train and look at the lake, and then the company would run them out to the irrigated lands in big Maxwell touring cars that were lined up at the curb. It was pretty effective. There's an awful lot of folks in these parts who can trace their roots to the Midwest." He paused for a drag on the cigarette. "I did a little archaeological dig for the City of Plainview a few years ago when they were refurbishing the depot, and I found they'd lined the edges of that old playa with concrete. It was covered up with sand and silt, had weeds growing in it. Been that way since I was a kid. But evidently they weren't taking any chances."

LAKE PLAINVIEW LASTED just five years, from 1912 to the day the pump broke down irrepairably in 1917. It was enough; groundwater irrigation had become firmly established in Texas and was spreading to other parts of the High Plains. The key was not so much Lake Plainview as it was a piece of new technology: the deep-well centrifugal pump.

Vaccuum pumps—mechanically speaking, the simplest kind—operate on air pressure, and are thus limited to the height to which air pressure can lift water: thirty-four feet at sea level, less at higher elevations. Reciprocating pumps, which use a pair of one-way valves to pass water through a cylinder into a pipe reaching to the surface, can go a lot deeper: as much as a thousand feet, if the valves are tight and the machinery can handle the weight of a column of water that tall. A reciprocating pump cannot, however, move water very fast. The limiting factors are the size of the cylinder and the speed at which the piston (in a single-action pump) or the cylinder itself (double action) moves back and forth. A cylinder with a one-gallon capacity moving back and forth twenty times per minute will move twenty gallons of water per minute—no more, no less. Larger and faster pumps can move more, of course, but there are limits. Too big, and the weight of the water becomes prohibitively heavy; too fast, and the reciprocating machinery stands a good chance of flying apart.

Centrifugal pumps have far fewer limitations. They pump by means of a vaned rotor, called an impeller, which spins rapidly inside a circular chamber. Water enters the chamber through a hole near the impeller's axis and is flung violently outward by the spinning vanes, hence the name "centrifugal." The water then passes into the delivery pipe through an opening on the outside of the chamber. Early centrifugal pumps depended on the vaccuum created inside the chamber by the whirling impeller, giving them the same height limitations as any other vaccuum pump, but it was quickly realized that they could push much better than they could pull. The impeller and its chamber were sunk to the bottom of the well, creating the deep-well pump. With a powerful-enough motor and a fast-enough rotor, a farmer could now lift just about any amount of water from just about any depth. The stage was set for a second technological innovation, one that would irredeemably alter the face and the future of the High Plains.

XV

CIRCLES ON THE PLAIN

B ECAUSE THE OGALLALA AQUIFER is both vast and
largely hidden, it is difficult to comprehend the threats that it
faces. Springs drying up in New Mexico seem unrelated to the
number of wells in Nebraska; falling water tables on the Llano Estacado have no meaningful physical connection to falling water tables in
southwest Kansas. Groundwater flows slowly: In the Ogallala, the rate
is about a foot per day. That is just over a mile in fifteen years, or sixty-
six miles in a millennium. At that speed, it is impossible for any one
well to impact more than the few square miles directly around it in
any meaningful way.

But it would be a mistake to assume that a lack of physical connections among the water problems of the High Plains means that there
are no connections among them at all. The ties that bind the far-flung
Ogallala's declining wells to one another are not tangible, but they are
nevertheless profound. They are rooted in the same human failings,
and they bode, ultimately, the same hard future. That future may not
take place exactly as we envision it. It will certainly not take place
everywhere over the aquifer at the same time. It will look different in
different regions, depending upon local soils, climate, topography,
population, and a host of other factors, not all of them now known.

But it is the same future. If we are to deal effectively with it, it is necessary to acknowledge that fact.

I have said that the Ogallala is hidden. That is not completely true. For those who understand what they are seeing, there are some big, out-in-the-open, in-your-face indicators that the underground ocean is present any place you happen to be on top of it. These visual attention-getters—at the same time markers, value-makers, and maledictions—are Frank Zybach's circles.

FRANK ZYBACH WAS a tinkerer. I don't mean he simply messed around with machinery; I mean he thought with his hands. Like many of that ilk, he was not fond of school and left it after completing the seventh grade. That was in 1906. His father put him to work on the family farm near Columbus, Nebraska, where the Loup River marries the Platte sixty miles upstream from the Iowa border. The elder Zybach assigned his son, then thirteen, the task of walking behind a horse-drawn harrow to prepare the fields for planting. That was the genesis of Frank Zybach's first invention: a small cart with wheels that swiveled to turn corners, but could be locked when moving in a straight line. Attached to the back of the harrow, it allowed its young inventor to sit instead of walking while he guided the horse.

By 1915 Zybach was working on his first patentable device, a guide with a runner that would allow a tractor to plow a field by itself. The trick was to plow in a spiral. The farmer was supposed to complete the outer ring of the spiral, attach the guide to the tractor, and place the runner at the beginning of the furrow. He could then go and do something else; the runner would follow the furrow and keep the driverless tractor on course all the way to the center of the field. Zybach obtained his patent in 1920 and sold the manufacturing rights to a farm-implement company in Lincoln. It fell flat. Tractors in those days were heavy, expensive, and, if uncontrolled, apt to run amok and destroy small buildings. Few farmers wanted to risk leaving one running unattended. The most interesting thing about that early patent, in the light of later events, was that it appears to have encouraged its inventor to think in circles.

The story of the tractor guide is the story of all of Frank Zybach's early inventions: interesting in concept, a bust in application. No money was made. By 1947, the fifty-three-year-old would-be inventor was living near Strasburg, Colorado, on the rolling Rocky Mountain piedmont forty miles east of Denver, just west of the raised edge of the High Plains. He had found work as a tenant farmer, plowing another man's fields because he could not afford to purchase his own.

IN THE 1940S, irrigation was still pretty much synonymous with flooded ground. Water flowed in ditches that ran across the tops of slightly sloped fields. The fields were plowed at right angles to the ditches; water released from the ditches would flow slowly down the furrows, wetting the soil as it went. In its crudest form, flood irrigation was controlled by nothing more complicated than a spade. A farmer would shovel openings in the ditch wall to let water onto his field; he would pack dirt into the openings again when he wanted to stop the flow. Later, boxes made of lathe were set into the ditch walls. Canvas dams placed across the ditches backed water up so that it would flood into the lathe boxes, and from there into the furrows; removing the dams would drop the water level below the lower lips of the boxes, and the flooding would stop.

In the late 1930s, farmers began to use plastic siphon tubes to move water from ditch to furrow. That eased their work and extended their control over the water, but the water was still coming from ditches and it was still flowing over the ground. There was evaporation from the ditches and from standing water on the fields, and there was runoff ("tailwater") escaping from the bottoms of the furrows. Irrigation farmers hold rights to certain fixed amounts of water, and to watch that water disappearing into thin air or onto a neighbors' property, not sinking into their soil and growing their crops, nagged at them. A few began to experiment with sprinklers.

Frank Zybach's rented farm had too much slope for furrow irrigation, so he was planting dryland wheat and not being very successful at it. A neighbor stopped by and suggested that they go together to a demonstration of one of the new sprinkler systems, which were supposed to be usable on sloping fields. Zybach left the machinery he had been cobbling together for one more stab at dryland cultivation and came along.

Watching the demonstration, the aging tinkerer found himself fascinated by the idea of irrigating with sprinklers, of watering the land with a gentle, controlled rain instead of drowning it like a cloudburst. He was less impressed with the sprinklers themselves. They were moved by hand: Each time you wanted to put water on a different spot in the field, you had to wade through the soggy muck around the sprinkler, disconnect all the hoses and fittings and pipes, lug everything to the new location, and set it all up again. It was messy, it was labor intensive, and it was time consuming. Zybach remarked to his neighbor that there had to be a better way.

Within a year, he had come up with one. Frank Zybach's version of the irrigation sprinkler didn't have to be moved from one spot to another; hearkening back, perhaps, to his tractor guide, it ran itself around in a circle. A vertical pipe rising out of the ground at the pivot delivered water through a rotating coupling to a horizontal line with sprinkler nozzles mounted every few feet; the horizontal line was held off the ground by wheeled towers. The water provided motive power as well as irrigation. As it passed through each tower, part of the stream was diverted through a double-throw cylinder; the cylinder drove a rotating arm that caught on a series of lugs on the rim of one of the tower's wheels, forcing the machine forward. The towers were powered independently. The outermost tower moved constantly; the rest were controlled by valves that would open as the line of towers bent, and then close—cutting off the drive water—when the line became straight once more.

By 1948, Zybach had a rough prototype in operation. By 1952, he had his first customer—a neighboring farmer named Ernest Engelbrecht, who put in a five-tower system designed to irrigate forty acres. Zybach also had a patent, number 2,604,359, issued on July 22, 1952. That number and that date are now engraved permanently in the memories of historians of agricultural technology, at least one of whom, William E. Splinter, has gone so far as to call the center-pivot sprinkler "the most significant mechanical innovation in agriculture since the replacement of draft animals by the tractor." It would take time, but Frank Zybach's circles would eventually change the face of farming, not just on the High Plains, but over the entire planet.

EARLY ON, IT APPEARED as though the Zybach Self-Propelled Sprinkling Apparatus would go the way of Frank Zybach's other inventions. In its first two years, the manufacturing firm he set up back in Columbus with a boyhood chum, auto dealer A. E. Trowbridge, sold just nineteen units. Part of the reason was reluctance on the part of farmers to try the newfangled, untested device; the rest of the reason was Frank Zybach. Ever the tinkerer, he insisted on making improvements to each new machine, shrugging off Trowbridge's pleas to leave the damn thing alone long enough to train people to build it and service it. The two friends found themselves bickering. To salvage their relationship, Trowbridge and Zybach sold the manufacturing rights to a fellow named Bob Daugherty, who owned a small shop that built balers, hay rakes, and silos in a little town called Valley, a few miles west of Omaha. Daughtery borrowed tool-up money from a local bank and set to work. The Valley Manufacturing Company sold its first center-pivot sprinkler in 1955 to a farmer in Holt County,

Nebraska. Foreshadowing things to come, that first sprinkler pumped water from the Ogallala Aquifer.

Valley Manufacturing sold just six more units the first year—fewer, even, than Zybach and Trowbridge had sold the year before. The second year wasn't much better. Daugherty pleaded with his banker to sit tight. He had faith that the sprinklers would catch on. He was right.

AKRON, COLORADO, IS a tiny town set barely onto the western edge of the High Plains. East and south, the land lies flat to the horizon; west and north, it steps down abruptly onto the broad alluvial plain of the South Platte River. The day I arrived, the town was seeing its first hard rain after a long dry spell, and the spring-green land was embracing it eagerly. There were puddles and there was mud and there were the scents of damp and growing foliage. The sun shone between shafts of falling raindrops.

"Hello! You brought the rain!" Wayne Shawcroft greeted me happily. An amiable man in his late sixties, Shawcroft is a banker. Before that career, though, he worked as an irrigation specialist at the agency where he had agreed to meet me—the U.S. Department of Agriculture's Central Great Plains Research Station, just outside Akron. "I had Joel's job," he explained, glancing over at Joel Schneecloth.

Schneecloth nodded. He was perhaps forty years old, loosely put together but intensely focused. "A guy on TV this morning," he deadpanned, "was outlining a number of signs that would signal the end of the drought. And, of course, rain was one of 'em." The merest hint of a smile. "But also an increase in overall cloudiness. We've had that for the past two or three months. So this rain is probably a signal that the drought is breaking. For us."

Breaking a drought has extra importance for Akron, which is one of the very few places over the Ogallala that has *not* gone heavily to center-pivot sprinklers. This close to its western edge, the aquifer begins to pinch out; there is not enough saturated thickness to support groundwater irrigation. It thickens quickly to the east. The pivots begin a few miles east of town; by the time you get to Yuma, thirty miles away, they are pretty much everywhere.

Shawcroft remembers the first center-pivot sprinkler in Yuma County. "You go straight north from Yuma, up to the first bend on Highway 59," he told me. "That's where it was. A guy developed a little well up there, and he was irrigating corn. That was in the late 1950s. And then along came the 1960s and boy, those center pivots just exploded over there. It was a complete change. All that area used to be dryland. They went over to the philosophy that they were irrigated country."

"Well, that was during a period of dry weather," pointed out Schneecloth.

"No, it wasn't that," Shawcroft protested. "It wasn't necessarily any drier. It was the invention of the center-pivot sprinkler that did it." He laughed. "That was the only time I wasn't here. I was away in graduate school from 1965 until the spring of 1970. And when I came back I was totally shocked, because Yuma had changed from the typical dry-land area we were used to around here to irrigated agriculture."

What would fuel such a rapid cultural shift—especially one involving the purchase of a complicated, relatively untested machine a quarter of a mile long? Putting in a center-pivot sprinkler requires a high initial capital outlay; the per-unit cost today is around forty-one thousand dollars. Was it water conservation that drove farmers? Easy credit? An itch to own the latest mechanical wonder?

Shawcroft wouldn't buy any of those reasons. "It was basically a labor situation," he told me. "The farmers didn't want to spend the time and energy to manage their furrow irrigation a little bit more intensely. They could have approached the sprinkler's efficiency through good management. But it was easier just to turn the switch on the center pivot."

The center-pivot sprinkler, which revolves around a field, was a revolution in other ways as well. It required less water and thus less fuel to pump the water—sprinkler installations surged during the gasoline shortage of 1973—but that was really a side issue. The big selling point was that tending a sprinkler was a lot less work than tending ditches and furrows. That meant you could irrigate significantly more land, and make significantly more profit.

Furrow irrigation did not stop evolving when sprinklers came in. Modern flood systems replace the old ditches and slat boxes and siphon tubes with gated pipe, reducing evaporative loss and greatly increasing manageability. The gates in the pipes can be set by computer, offering precise control over how much water is released and when it is placed on the ground. Tailwater is collected and recycled. A recent innovation, the surge system, wets the furrow with an initial flood of water, then reduces the flow once the wetted earth begins to absorb water more rapidly. With these tools, furrow irrigation can approach the efficiency of sprinklers. It is also significantly less capital-intensive; a computerized surge system, the most expensive type of furrow irrigation, runs roughly three-quarters the cost of a center pivot. The center pivot, however, remains easier to operate and to maintain.

Today, Frank Zybach's circles mark the Ogallala as ice marks Antarctica or tides mark the sea. They are used widely in other parts of the world as well—in Australia, India, North Africa, Spain, even in damp England, where there is a cluster of Valley sprinklers just outside London, near Heathrow Airport. Bob Daugherty's little Valley Manufacturing Company is now Valmont Corporation, with a sales force in nearly every country in the world. Numerous competitors have sprung up; much money has changed hands. Zybach and Trowbridge, who were shrewd enough to negotiate a 5 percent royalty on every unit sold, made well over a million dollars each before they died, within a few months of each other, in 1980.

By that time, the doubts had begun. Wells were drying up in southwest Kansas and in the Texas Panhandle; water levels were dropping with alarming speed in Oklahoma, Colorado, and New Mexico. Crops were failing; pumping costs were climbing. The Ogallala Aquifer had once appeared bottomless. Suddenly, the bottom seemed uncomfortably close.

XVI

APOCALYPSE DELAYED

ONCERN SURFACED AS early as 1953. In a U.S. Geological Survey groundwater bulletin published that year, geologist Stan Lohman noted that withdrawals were already exceeding recharge over most of the southern part of the Ogallala, and that water tables in New Mexico and Texas were beginning an alarmingly steep decline. Pointing to the aquifer's importance as the sole water source for most of the High Plains, Lohman called for a study to determine such basic information as recharge rates, discharge rates, and total water in storage—all still unknown, though the Ogallala had been on the maps for more than fifty years.

Lohman's concerns were reiterated a decade later by Wayne Clyma and Fred Lotspeich of the U.S. Department of Agriculture's Research Station at Bushland, Texas. In a paper presented to the winter 1963 meeting of the American Society of Agricultural Engineers, Clyma and Lotspeich argued that the decline Lohman warned about had since multiplied enormously:

Approximately 10 percent of the original volume [of water in the aquifer] had been depleted by 1958, with the major depletion occurring since 1950. . . . South of the Canadian River and in Texas,

20 percent of the area had been seriously affected by declining water levels and 64 percent of the area had been affected moderately.

There was little question about the cause. The two Texans included a graph that showed the number of irrigation wells present on the Llano Estacado each year of the previous decade. In 1950, there had been fewer than ten thousand. Just ten years later, the number had surged to over fifty thousand.

The problem with early warnings such as those sounded by Lohman, Clyma, and Lotspeich was that they occurred in papers written by scientists for other scientists. The language was as dry as the topic, and interest was limited to those who were already interested. That changed in 1979. In February of that year, the *Wichita Eagle and Beacon* published a series of articles under the title *We're Running Out*, and the nonscientific public finally woke up.

The *Eagle and Beacon*'s language was pretty blue. "Today, the Ogallala, like a great rock sponge, is being wrung dry from three decades of continuous irrigation," lamented authors Karen Freiberg and Martha Mangelsdorf. "State water resource experts predict that irrigation will be nothing but a memory in many large areas of west central Kansas in eight to ten years." Freiberg and Mangelsdorf tied the declines directly and explicitly to the center-pivot sprinkler:

> Like giant metallic insects, the steel-spined irrigation systems seem rooted in the western Kansas countryside. Whether they stretch above the corn stalks and milo stubble or burrow hundreds of feet into the clays and sands beneath, their mouths are anchored to a common prey—the Ogallala aquifer.

In thirty years, Frank Zybach's circles had gone from concept to struggling reality to ubiquity to, in the minds of growing numbers of High Plains residents, scourge.

———

IN THE SUMMER OF 1984, a major symposium on the Ogallala's problems was convened at Texas Tech University in Lubbock. More than thirty lectures were presented in eight broad topic areas. Symposium chairman Robert M. Sweazy, the director of Texas Tech's Water Resources Center, tried to keep things positive: He called his opening remarks "Recharge, Reuse and Recovery" and concluded them by noting that "tremendous strides" had been made toward Ogallala conservation. The presenters who followed him were considerably less optimistic. David Aiken, a water-law specialist from the University of Nebraska, was particularly blunt in his remarks concerning the symposium's host state:

> Texas . . . has followed the politically convenient local control approach to groundwater management. The result is virtual economic depletion of irrigation groundwater supplies in the Texas High Plains by the end of the century and the associated reduction of agricultural and related economic activity. High Plains irrigators have thus far attempted in vain to obtain a source of supplemental water to rescue themselves from their failure to control groundwater depletion. . . . Texas serves as an example to other states concerned with groundwater depletion; not of how depletion should be controlled, but rather what will happen when depletion is not controlled.

The first item on the symposium's agenda, a presentation with the modest title "The High Plains Regional Aquifer—Geohydrology" by a pair of geologists named Ed Gutentag and John Weeks, drew an overflow crowd. This was due not to the paper's timing, but to its authorship: Gutentag and Weeks were coordinating a USGS team that was conducting a massive ten-year study of the aquifer, and this was their midterm report. Authorized by Congress in 1978, the High Plains

RASA—Regional Aquifer-System Analysis—was unprecedented in several ways. One was the scope of the material being analyzed, which included historical accounts, previous geological studies, drillers' logs and water-level data gathered from more than twenty thousand wells, stream gauges and spring-flow measurements, weather and climate records, and LANDSAT images. A second aspect was the geographic breadth of the study, which ignored political and watershed boundaries to take the first really uniform look at the whole of the High Plains. The most important difference from previous studies, though, was the method of data analysis. Earlier attempts had been made to create computer models of local regions of the Ogallala. Gutentag, Weeks, and their colleagues created a model for all of it.

Actually, they created three interconnected models, one each for the northern, central, and southern High Plains. The regional divisions were set at the two points where the aquifer becomes wasp-waisted—Palo Duro Canyon near Amarillo, Texas, and at the headwaters of the Smoky Hill River on the Kansas–Colorado border. The models were very rough—the datum points (the locations where measurements are taken to be fed into a model) were ten miles apart—but they gave a much better sense of the dynamics of water within the aquifer than anything that had been attempted before.

The picture that emerged was not pretty. Here is part of what Gutentag and Weeks reported in Lubbock:

> In most of the areas being irrigated in the High Plains, ground-water is being mined; that is, more water is removed annually from storage in the aquifer than is replaced by recharge. Areas of decline caused by irrigation pumpage are found in all states except South Dakota. . . . The areal extent of these declines has increased considerably during the last 30 years and the maximum decline has increased from about 45 feet to nearly 200 feet.

This language was repeated almost verbatim four years later in the eight-volume final report of the study, where it was accompanied by thickets of statistics. One of the more alarming trends highlighted by those (literally) dry numbers was a dramatic drop in the acreage irrigated by each well. The declining water table had decreased hydraulic pressure within the aquifer, causing well yield to shrink. In Texas, this had slashed the acreage the average well was able to serve by almost 50 percent, from 118 acres to 62, in less than thirty years.

READING THE HIGH PLAINS RASA TODAY, one emerges with a strangely mingled sense of alarm and relief. The alarm stems from the picture the report paints of a resource in severe decline; the relief arises from the knowledge that the decline has not continued, at least not at the rate that Gutentag and Weeks predicted it might. The apocalypse has failed to arrive. Water tables and well yields have continued to drop, but at slower rates than before. Agriculture remains viable over large areas of the High Plains that were supposed to have run dry by now.

To some, this suggests, if not a flaw in the study, at least a flaw in the reporting of the study. Jim Goeke, the Nebraska geologist, is one of those. "I really think you can create a bogeyman," he insists, "and I think the High Plains RASA did. It just scared everybody to death, because the projections were commonly for fifty to one hundred feet of decline above and beyond what was already there. Those declines haven't materialized, for a number of reasons. If you read the current studies, they talk about changes in aquifer character, and things like that. But often what you're really talking about, more than aquifer changes, are cultural changes. Those have been profound. It's the cultural practices, I think, that have had more impact than anything you can tie directly to the Ogallala."

Goeke is at least partially correct. There is no doubt that the culture of water on the High Plains has altered significantly in the fifteen years since Gutentag and Weeks issued their predictions. Much of this has come in the form of improvements to water delivery: Less is wasted between the well and the crop. Tillage practices—the preparation of the ground for planting, and the care of it while it lies fallow—have also changed, in ways that improve the land's ability to retain the water that the improved irrigation practices put onto it. Perhaps most important, attitudes have changed. There is now near-universal agreement that the water in the Ogallala is a limited resource that must be used sparingly. Wasting it is not just carelessness, it is a social error. As one county agent in Texas told me, "If you do that now, people will get offended real fast."

But there are at least two other sides to the current view of the High Plains RASA. First, it is necessary to understand that many of the changes that have nullified the report's more draconian predictions were brought about by the report itself: Concerns about the future of the aquifer were widely accepted only when it was no longer possible to deny them, and those concerns have led directly to the improvements that have slowed the aquifer's decline. Second, the reprieve brought about by cultural change is, as we shall see, likely to be only temporary. The apocolypse has only been delayed, not prevented; the decline could resume its old, alarming speed at any time.

THREE

MINIMUM

WATER

XVII

ADAPTATION

O N SEPTEMBER 11, 2001, in the wake of the terrorist strikes on the World Trade Center and the Pentagon, the Federal Aviation Administration closed all airspace over the United States and pilots of all planes in the air began looking for safe places to land. Fear of further attacks caused traffic to be diverted from large metropolitan airports, so commercial flights had to search out smaller fields with runways long enough to handle large aircraft. Three of them came down at Garden City, Kansas. Lacking moveable ramps and portable stairways big enough for jumbo jets, the airport phoned the Garden City fire department, which sent a hook-and-ladder truck onto the tarmac to disembark the passengers.

"This is a small airport," remarks Mahbub Alam as he drives us past, "but it was very important on September 11, when the air space was closed." After more than twenty years in this country, his words still carry the soft, clipped accent of his native Bangladesh. "People from all over the world were stranded in Garden City. You could come here, and you could see all the big jetliners—I never thought such big planes could ever land here."

Mahbub Alam is a small, graying man with a meticulously trimmed moustache, wire-rimmed photogray spectacles, a neat olive-drab shirt, and an oversized purple baseball cap with K-state inscribed above the bill. The incongruous cap is really a badge. Alam is the Kansas State Extension Service irrigation engineer for western Kansas, serving a fifty-county district out of an office in Garden City. Sixty-two years of living, too many of them at a desk, have stiffened his joints: Following him around the extension service campus on East Mary Road a few minutes earlier, I had noticed that he rolled slightly in his Nikes, like a sailboat on a barely windy day.

The airport is at the east edge of Garden City, on flat land, which is the only kind of land they have around here. It is besieged by irrigated fields. North, east, and south, the center pivots stretch away to the horizon. April is in the air, and the circles of dark earth beneath the sprinklers are just beginning to turn green. Meadowlarks carol beneath an Ogallala blue sky.

In 1943, with war raging in Europe, the Army Air Corps came to Kansas looking for a place to build a training base. They chose a fifteen-hundred-acre site near Garden City, where the land was level and available and the closest invasion route by sea was nearly a thousand miles away. The Garden City Army Airfield had three squadron hangers, five runways, a personnel roster of thirty-five hundred, and as many as fourteen hundred aircraft at one time. When the war ended, all of those moved out, and Garden City inherited the base. That is how this small Kansas town came to have an airport with a runway long enough to handle a 747. It probably could not be duplicated today. Back in 1943 there was no such thing as a center-pivot sprinkler, the ditches from the Arkansas River were as dry as the river itself, fuel for groundwater pumping was dear, and land around Garden City was dirt cheap. Today, the city is at the heart of irrigation on the central High Plains, some of the most productive agricultural land in the world. The fields lap greedily at the airport, as if wishing to swallow it. Had the lumbering, propeller-driven bombers of sixty years ago not needed all that space, much of it would likely be growing corn and alfalfa today.

Leaving the airport behind, Alam turns onto a paved farm
road, headed east. Center pivots line the road, moving slowly, like
giant, arthritic insects. We stop near one to examine it closely.

Irrigation sprinklers have changed dramatically in the nearly sixty
years since Frank Zybach attended a demonstration of a hand-moved
version and came home convinced that there had to be a better way.
Zybach's first model had a pair of towers ten feet tall holding a
pipeline eighteen inches off the ground; guy wires from the towers
suspended the pipe in a parabolic embrace, as the deck is suspended
on the Golden Gate Bridge. Later versions moved the pipe to the tops
of the towers in order to clear tall corn. On these, the pipe was either
hung from a rigid rod or bowed slightly upward and kept under tension by cables stretched between the towers. Nozzles on the pipe
jerked powerful streams of water around in circles, like huge institu-
tional lawn sprinklers. Such "high-impact" devices are not often used
today. Driving near Ingalls much later, Alam spotted one and stopped,
thinking we might be able to see it in operation. That was before we
noticed the small cedar growing between towers five and six, tenderly
nestling the long-defunct sprinkler pipe in its upper branches.

Up close, a modern center pivot resembles a lawn sprinkler less than
it does a medieval siege device. The one we have stopped beside near
the Garden City airport is typical. Tall triangular towers riding on
small wheels extend in a straight line from the road's edge to the field's
distant center. The towers support a pipeline seven inches in diameter,
fifteen feet off the ground, and a quarter of a mile long. The pipeline
sprouts nipples every five feet; attached to the nipples are drop tubes,
black plastic tubes an inch through and twelve feet long. The drop
tubes end in hanging nozzles from which emerge circular patterns of
water, like the spokes of wheels. Rainbows pirouette in the spray.

As we watch, the outermost tower lumbers briefly to life, creeps forward a few feet, then stops. Down the pipe toward the distant pivot

the other towers fall obediently into line like well-drilled cheerleaders. Like most current sprinklers, this one is electrically powered. The outer tower is controlled by a timer; the motors of the rest are turned on and off by contact switches mounted on the pipe, the electrical equivalents of Frank Zybach's water-drive valves. As the line bends, the switches close, setting the towers in motion until the pipe is straight enough that contact can no longer be maintained. Center-pivot timers are calibrated in percent per day, rather than in feet per minute or miles per hour. Judging by the timing of the towers' episodic motion and the distance moved during each episode, this one was probably set at 20 percent. That would take it once around the field in approximately five days.

The outer towers inscribe much larger circles than the inner ones do, so the amount of water each nozzle must supply increases substantially as you move away from the pivot. This is accomplished through a combination of pressure regulators, which control the water pressure individually for each nozzle, and the size of nozzle orifices. Drop-tube nozzles come in graduated sets that increase in size in a stepwise manner from the pivot out; each nozzle carries a number indicating which step it belongs to. The numbers change at the towers. Inattention to this detail can cause problems. Several times on the High Plains I heard stories of farmers who had lived with diminished crop yields—sometimes for many years—because they had read the numbers wrong, or because they had accidentally switched the numbers on two adjacent lengths of sprinkler pipe, or because of what Alam calls "creeping maintenance"—coming across a broken nozzle during an inspection and replacing it with a new one of the wrong size that happens to be kicking around in the back of the pickup. Mistakes such as these are responsible for a small but noticeable portion of water problems on the High Plains.

The nozzles come in different configurations as well as different sizes. All share the same basic design: water is shot downward onto the center of a round plate roughly an inch across, which breaks up the stream and spreads it out in a circular pattern. But small differences in detail can make large differences in performance. The plates may be flat or grooved, narrow or broad, concave or convex. Some plates are stationary; others move when the water hits them. Wobblers rock back and forth, simulating slightly uneven rainfall; trashbusters, designed to be used when irrigating with runoff from animal feedlots, have grooves with a steep slant to prevent hair and particles of manure from clinging to them. Spinners are driven by small electric motors and can throw water as far as forty-five feet. The spray may spread in an even circle beneath the nozzle, or it may form a doughnut, with the nozzle at the center of the hole. A few designs are one-sided: These are for use near the towers, to keep mud from developing in the paths where the wheels run.

Almost all current sprinklers put the nozzles at the ends of drop tubes. These conserve water, but that is not why they have been so widely adopted. The primary impetus behind the switch from high-impact sprinklers to the more efficient drop-tube versions has not been conservation for the future so much as it has been adaptation to the present. As the water level has declined in the Ogallala Aquifer, the yield from each well has declined with it. Yield determines the amount of pressure a well can sustain while running a sprinkler system. High-impact systems require fifty to one hundred pounds per square inch; drop tubes can get by on ten to fifteen. In their extreme form—the ultra-drop system known as low-energy precision application, or LEPA, which places the water directly on the ground—they can survive on as little as four.

LEPA tubes end in applicators, rather than nozzles. These are of two types. One type, which drifts along a few inches above the ground, has an opening that forms a bubble rather than a spray pattern; water is delivered to the soil at the point where the bubble touches the furrow. The other type is a permeable cloth sock that is dragged along the ground behind the moving sprinkler, leaking water as it goes. Proponents claim that with a properly run LEPA system as much as 98 percent of the water applied by the sprinkler gets to the crop. That compares to 85 percent for conventional drop-tube sprinklers, 75 percent for high-impacts, and as little as 50 percent for furrow irrigation.

It was a Texas farm boy named Bill Lyle who first came up with the idea for a sprinkler that would lay the water directly in the furrow rather than spraying it into the air. Lyle was studying agriculture at Texas Tech University in Lubbock, a half-hour west of the family farm just outside the little town of Ralls. This was during the early 1960s, and center pivots were rapidly replacing furrow irrigation on the Llano Estacado. Making the drive between school and home, Lyle watched the wind carry spray from the sprinklers onto neighboring fields and thought, like Frank Zybach before him, that there had to be a better way.

"I wanted to come up with a way to spoon-feed water to a crop," he told an interviewer for *Lifescapes* magazine forty years later. "Give it only what it needs and put it right down on the ground between furrows to eliminate evaporation and drift."

Lyle eventually ended up with a Ph.D. in agricultural engineering and a job with the Lubbock District of Texas A&M University's Research and Extension Service. It was there, in a cluster of low, beige-colored brick buildings on a road designated FM 1294, just north of the Lubbock International Airport, that the first LEPA system was put together. The team that assembled it consisted of three people: Lyle; Jim Bordovsky, now with the district's Halfway research facility, near Plainview; and Leon New.

LEON NEW IS A TINY, courtly, white-haired Texas leprechaun whose nonstop monologues about LEPA are sweetened by a honey-rich southwestern drawl. His office in Amarillo, where he moved in

1983, is crowded with the organized clutter of an individual who is not inherently messy, just too busy to put things away. Numerous plaques and awards line the walls; irrigation nozzles and LEPA applicators perch on stacks of paper on the tops of file cabinets. A bundle of drop tubes leans in one corner.

"We brought the water down to the ground," he tells me. "With the impact sprinklers, we 'uz throwin' water as high and far as we could throw it. We came from the high-impacts in the nineteen-seventies to right over the canopy—we called that MESA, mid-elevation spray application. Then we lowered that further. We used a one-liter Pepsi bottle to get the water onto the ground—just turned it upside down and put a connector in the bottom, and the water came out that little throat. We used a little bit of everything, see, to get this principle across."

New's role was field testing—finding farmers who were willing to be guinea pigs, then building and monitoring LEPA systems on their farms. "The growers are the people who made it work," he emphasizes. "They would let me go on their fields, and I tried to plan it to where we wouldn't bankrupt them. These were people who believed in what we were doing. I've had growers say, 'Well, we just kind of tolerated you when you were doing all this, but we're glad you did it.' Because of the water in the Ogallala, see, and the way it's going."

As the farmers used the tested systems, the LEPA team tweaked them to make them more farmer-friendly. Putting drop tubes in alternate furrows instead of every furrow reduced water use and cut the costs of conversion in half; speeding up the pivots improved water absorption and decreased runoff. The Pepsi bottle was replaced by a convertible applicator that could create either a LEPA bubble or a spray—critical for chemigation, and for propagating plants, such as wheat, whose seeds will not germinate unless they get wet.

A crucial factor turned out to be the drop tubes themselves. "One of the first systems that went full-scale was a half-miler," New recalls, smiling. "We had twenty-eight hundred gallons a minute in it, and

we'd moved the pressure regulators off the bar and down to the applicators. So the farmer calls me one day—it was about a hundred and
two degrees outside—and he says, 'The hose is ballooning!' We were
running such small amounts of water in that hundred-degree weather,
see, we couldn't cool the hose, and it just bulged out. If it would of
blown, he'd of had a mess. Hundred-and-two, hundred-and-three
degree weather in corn, and you don't have time to stop." The next
batch of drop tubes was made from stronger hose.

LEPA is a complete system of farming, not simply a new type of
sprinkler. The long drop tubes reach below the furrow lips, which
means that fields must be plowed and planted in circles to avoid hanging up the applicators as they cross the rows. Furrows must be
dammed and diked to contain the runoff caused by concentrating the
water. Fields are often laser-leveled to improve uniformity of application. Operating costs climb. These are among the prices we pay to
maintain our dependence upon a declining resource.

Not everyone believes the prices are worth it. One irrigation engineer I spoke with in Nebraska scornfully referred to LEPA as "moving
flood irrigation." He described a test he had run on a field with "a little
bit of a slope. A *little* bit of a slope—1 percent, maybe 2 percent. We
had water running three hundred feet in front of the system." Adjusting the pivot speed can help that, but it cannot help the fact that concentrating water in the furrows leaves the ridges dry. The uneven
moisture pattern that results concentrates plant roots instead of
spreading them through the soil: it is simply no match for the even,
gentle, artificial rain of a sprinkler.

Farmers have complained that harrows, mechanical planters, and
other farm equipment designed to cover multiple furrows cannot turn
tightly enough to be used toward the center of a LEPA circle. You cannot drive a truck across a circular-plowed field without jolting across
furrows, damaging furrow walls and increasing wear and tear on the
truck. These things can be dealt with, but only through greater

expense and greater labor. Farmers who dropped their operating costs and their labor by going to center pivots are understandably reluctant to jack them up again just to save a little water. LEPA undoubtedly has a place, in some regions, with some crops. It is not a cure-all for water table decline on the High Plains.

And LEPA is only the penultimate step, anyway. There is one more level of improvement possible in the game of saving water through technology. After fifty years of intensive irrigation on the High Plains, there are many wells that can no longer support sprinklers—even sprinklers with LEPA applicators. For these, only one technological option remains: subsurface drip irrigation, or SDI.

WE HAVE LEFT THE PAVEMENT and are bumping slowly down a narrow dirt road a dozen miles east of the Garden City airport next to a healthy field of dark green alfalfa in which no sign of conventional irrigation can be seen—no damp furrows, no ditches, no sprinkler. Weeds growing between the tire ruts tap dance on the undercarriage. Large white boxes with slanted lids stand beside the road every hundred feet or so, their backs to us, like bashful, overgrown beehives.

"This used to be sprinkler," Alam comments, patiently negotiating the ruts. "He took the sprinkler out because his well capacity went down, and now he is irrigating with drip. He first put it in for corn, and he did raise good corn. But the corn price is not that good, so he went into alfalfa. This is pretty interesting, because he has only a 240-gallons-per-minute system. His application is one-tenth of an inch per day. That is all he can apply."

The owner of the drip system arrives precipitously a few minutes later, at the wheel of a small flatbed truck. Dave Wehkamp is a big, carelessly muscled young man in a khaki T-shirt and blue jeans; his belt sports an oval buckle nearly as broad as the field we stand beside. "I'd've shaved this morning, Mahbub, if I'd known you was gonna bring visitors," he complains, rubbing at least two days' worth of stubble.

"You look great," Alam assures him. "You look like a Hollywood actor."

Wehkamp smiles wanly. "I've just been running awful hard."

"It's a big challenge to take, I know."

"It has been. It has been."

I ask Wehkamp why he put in the drip system. "Most of it was economics," he responds. "The well was small enough to where I didn't have enough water, I didn't feel, to justify a sprinkler. It was just a better money-making venture for me to do it this way, even though it was a little more expensive per acre."

"What made you decide for drip over a LEPA system? They're almost as efficient—at least, that's what you hear."

"That's what you hear," Wehkamp agrees, cautiously. "I'm not convinced of those numbers, I guess. I don't actually figure there's any difference, for me, between LEPA and a sprinkler. On this hay ground, I develop a hardpan, so with either one I can't get the water in the ground." The drip system sidesteps the hardpan; it places the water directly in the soil, in the root zone of the plants. In its brief sojourn from underground to underground, it never sees the sun.

"Are you getting the payoff you expected?"

"I thought the labor would be less than what it is. But the payoff is about what I expected."

"So you feel it was worth doing?"

Wehkamp hesitates, then laughs. I nod toward Alam, who has wandered off and is digging through the car for his camera. "Mahbub isn't within earshot right now," I point out, "so you can answer."

Wehkamp laughs again. "In this instance, I do," he says. "But, you know, I've got plenty of other ground where I can't make it work."

"So you're still using sprinklers on that."

"Oh, yeah. Again, a lot of the reason is economics. It dictates that. I'm still growing a good enough crop on those fields, because I have more water. This one was either a dryland situation or drip. And once you have the investment, in the well and the pump and stuff, you've gotta look."

He lifts the lid off the nearest of the white boxes and explains the operation of the system. It is straightforward. Individual water lines lead from the pumphouse to each box: at the boxes, the lines split into Ts buried four feet deep along the edge of the field. Risers along the branches of the Ts feed drip tapes—flexible flat water lines with tiny holes in them—located fifteen inches beneath the furrows. Solenoid-controlled valves portion the water out independently to the two sides of each T; there are eight Ts, dividing the field into sixteen individually-tailorable zones. The pumphouse contains a chemigation system that can inject fertilizers or other agricultural chemicals directly into the main water line. A second set of injectors controls the acid-alkaline balance: Wehkamp is trying to hold it at pH 3.5. On the logarithmic pH scale, where 7 is neutral, 3.5 would be moderately acidic—about the same as a good red wine.

"Lowering the pH means there will be less encrustation in the drip tapes," Alam explains. "General maintenance is an issue with drip, because you want to extend the life of the system. The economics are better when you have at least fifteen years of life."

BACK IN THE CAR, driving away from Wehkamp's farm, Alam marvels at the drip system's ability to compensate for low well pressure. "Dave has very, very minimum water," he observes. "Most of the time it is point-one inch per day, whereas alfalfa will normally demand anything from point-two-five to point-three-five. He can apply not even half of that, but still it grows. It looks good. Still he is having a crop."

He hesitates for a moment, watching the road. "These are not all researched as yet," he says, finally. "But my feeling is that when you are irrigating with very low water, frequency may play a role. In his case, I think, he is seeing that. He is only applying a little amount, but with the drip he can apply it every day."

"It was interesting to me," I state, "that economics was his reason to put in an expensive system."

Alam nods. "Our studies show that when you are talking about a field of 160 acres—a 130-acre circle, the rest is corners—that is an area where the sprinkler has still advantage. But when you are talking about half a circle, in that case your advantage becomes with the drip. You can irrigate all eighty acres with the drip, whereas with the pivot you will be losing sixteen to twenty acres in the corners. Lifespan is also part of it. Our analysis was based on a life of twenty-five years for sprinkler and fifteen years for drip. If the drip can make a little more years, then it will become more favorable."

Alam is referring to maintenance issues: rodent damage, pump wear, encrusted emitters, rotting drip tapes. But there is another reason why a drip system may not last fifteen years: If water levels in the Ogallala continue to decline, the well may no longer supply enough pressure, even, for drip. The economics of water-saving irrigation equipment—the loans, the amortization tables, making the payments—all depend on having enough water down there for the water-saving equipment to save.

And here a paradox emerges. Conventional wisdom states that conserving water is a matter of reducing mechanical loss, that deploying more efficient irrigation equipment—drop tubes, LEPA, subsurface drip—will halt waste and save the Ogallala. But waste is not always waste, and saving is not always saving. Efficiency can be deceptive; it can doom the Ogallala instead of preserving it. There is a growing consensus among water people on the High Plains that this may be precisely what is happening.

XVIII

THE OPPORTUNITY TO BE EFFICIENT

SCENDING THE NORTH PLATTE RIVER through what is
now western Nebraska, wagon trains bound for Oregon
began to encounter strange, striking scenery. Tall buttes and
spires rose along the left side of the valley, eliciting fanciful names:
Chimney Rock, Jail Rock, Steamboat Rock, Dumpling Hill. Soon a
great wall of stone reached out to the river and blocked their path,
forcing the wagons over an arduous pass to the south. This was Scotts
Bluff, named for a fur trader, Hiram Scott, who died at a spring near
its base in 1828.

Today the twin towns of Scottsbluff and Gering sprawl at the base
of the bluff, Scottsbluff on the north bank of the river, Gering on the
south. The towns are tidy and prosperous-looking, with tree-lined
residential streets and downtown buildings of warm-colored Arikaree
sandstone, quarried from the same formation that Nature used to
build the bluff. Around the towns spread irrigated fields. Here in the
valley bottom the water comes mostly from the North Platte, but if
you climb out of the valley to the north or the south, you will be back
in center-pivot country. The High Plains stretch limitlessly outward;
the Ogallala Aquifer lies beneath your feet. Separated from each other
by bluffs, badlands, irrigation styles, and eight hundred feet of eleva-

tion, the plains farms and the riverside farms appear to occupy distinct and disconnected worlds. But the disconnect is not so great as it appears.

Conventional wisdom is full of advice about the Ogallala. Invest in water-saving technology: drop tubes, LEPA, drip. Stop sending water down unlined, leaky ditches. Stop throwing it in the air with high-impact sprinklers. Stop. Pump only enough for the crop's immediate needs: every drop beyond that is wasted. It all sounds eminently reasonable. Until you look at the North Platte River.

"THE CANALS IN THIS VALLEY are unlined, and they lose maybe 50 percent of their water," states Dean Yonts bluntly. A few years ago, if a statement like that had been spoken by a water professional—Yonts is the extension service irrigation specialist for the Nebraska Panhandle, a job equivalent to Mahbub Alam's in Kansas—it would have been followed immediately by a demand to line the canals. Today, it is going to lead us in a somewhat different direction.

"Right now, if you drive up to Guernsey, Wyoming, where they have the last of the dams coming down the North Platte, you can walk across the riverbed," Yonts continues. "The reservoir is shut off. But if you drive across the bridges here in Scottsbluff, you'll see water flowing in the river. That is return flow coming back from the canals. I don't know how long it takes—maybe three or four years—but it is water that has percolated down and come back in."

"Dean's right," agrees geologist Steve Sibray, who has joined the conversation in Yonts's office. "We have trout in streams in this area that would not exist unless they were getting water back from irrigation. If we were to line the canals in this valley, we would dry up the river."

The sprinklers on the plains above the valley's rim are part of the same pattern. An overwatering sprinkler, like a leaking canal, creates deep percolation—water draining into the lower soil profile, below the root zone of the crop. Deep percolation was once treated as a dead loss. But the water never actually goes away, it just goes down. Some will eventually return to the aquifer as recharge; the rest will crop out in gullies, as springs. The springs feed tributaries; the tributaries join the river. Overpumping the Ogallala's water is lowering the water table in the North Platte River watershed, but it is also helping to keep the river alive.

Conventional wisdom states that sprinklers lose water through evaporation and drift. The small droplets of spray evaporate rapidly in the heat of the day; they also get caught by the wind and end up on the field you're fallowing next door, or on the highway, or on your neighbor's pasture. That seems such an obvious point that it wasn't even questioned until recently. "It was often, 'Oh, my goodness, we're losing 30 percent of our water,' on that windy day, or that hot day," says Yonts. "But what they're actually finding is that the losses are not that high. The parts that evaporate and drift are the very fine droplets. The bigger droplets, which constitute the bulk of the water, fall down through. So what they're saying now is that by moving your sprinklers from above the canopy to down in the canopy you save on the order of only 1 to 3 percent of the water.

"People say, 'Look how much I can save by going to drip.' And there are some savings there, because you're not putting water on the surface. Early in the season, especially, when the plants are short and you've got all that empty ground out there, evaporation loss really is high. Late in the season, though, I don't think you'll see much savings between subsurface drip and sprinklers. Or even furrow. Because the only water that's truly taken out of the system is water that's consumed by the crop or evaporated from the soil."

"It's really a consumptive-use issue," Sibray adds. "You can convert, and become more efficient in irrigating, but the water that gets past the root zone will always end up back in the return flow system. So when you increase irrigation efficiency, you don't necessarily save water. An individual may save, but you have to take a whole-system approach."

THE LACK OF WHOLE-SYSTEM water saving is one reason to wonder about the value of water-saving technologies. Wayne Bossert gave me another. A tall, soft-spoken, introspective man who has been the manager of Kansas's Groundwater Management District #4 since its inception, Bossert was one of those quoted in the 1979 *Wichita Eagle and Beacon* series that many credit with first bringing public attention to the plight of the aquifer. Back then, he was heavily promoting water conservation. "Our goal is to cushion the return to dryland farming," he told the *Eagle and Beacon*'s reporters. "We're pumping at seven times the recharge rate. With no action, we could pump the water out in ten years at the rate we're going."

A quarter of a century after that quote was published, the water is still there. So is Wayne Bossert, but he is singing a slightly different tune. "Irrigation efficiency's kind of an intriguing concept," he told me in his Colby office. "And I've got mixed emotions, having watched it for so long. I'm not convinced that improving irrigation efficiency does as much for the resource as people think. That's because it's a capitalization back into the system. If a farmer spends a lot of money to improve efficiency, it almost necessitates that production go up if he wants to recover his investment." More crop production requires more water, cancelling out—and often reversing—the gains from increased efficiency. "It sounds good. We always push that—you can't *not* say you should be efficient—but in terms of solving the problem, it's less a piece of the puzzle than I used to think it was," Bossert continued. "Increased efficiency, if you put other constraints on it, is an excellent partial solution. But those extra constraints are pretty constraining. It can be done, but so far it hasn't, traditionally, *been* done."

I told Bossert about Dave Wehkamp's drip operation. He looked thoughtful. "We've got an operator out here west of town about four miles—just an unbelievable man, as a businessman" he said. "He was in a situation where he had about a 260-gallon-per-minute well, and he'd been trying to flood eighty acres with that. At the top of the field it was pretty good, but then the water dropped off, so the bottom third of the field, roughly, was difficult to irrigate with a flood system. He really wasn't even doing it; he was only watering the top forty acres for the last whole bunch of years. With that same 260-gallon well today he's gone to eighty acres of drip. Cadillac deal—I'll bet it cost him twelve hundred dollars an acre to put that stuff in. He's tripled plant populations—put all these micronutrients in—and he's using more water than he ever did before.

"It's better use of the water," he added quickly. "I'm clearly convinced of that. It's better use of the water. But it's *more water*. I think we need to talk about water use efficiency, from a management standpoint, as the same production with less water rather than more production with the same water. There's a wealth of improvements available in efficiency. But the producer is always going to go toward that other definition. That's the business sense of it all."

In the end, water table drop will be controlled, not by the efficiency of water use, but by how much water is pumped out of the aquifer. When well yield drops to the point where a high-impact sprinkler can no longer operate, a farmer has two choices: convert to a more efficient water use or shut the well down entirely. Shutting the well down will leave more water in the aquifer than pumping it, no matter how efficient the equipment that distributes it becomes. Water-saving equipment can dramatically extend the useful life of a well, but it will draw the Ogallala down farther than it would if farmers simply shut off the high-impacts when well yields become too low to support them.

"As far as conservation goes," Dave Wehkamp told me, speaking of his own drip system in Kansas, "it depends on how you look at it. If conservation means you're growing more crop, more benefit to the world, that's happening. But if you look at it that I'm using less water than my neighbor with a sprinkler, I'm not."

"The same water," agreed Mahbub Alam, "because the crop water need remains the same. Technology offers you the opportunity to be efficient, so you can stop wasting the water. If you want to increase the acreage, and double your production, that's a different question. Because then you are consuming more, definitely, because you are producing more. To produce more, you need more water."

That is, as Bossert says, the business sense of it all. If you cut your water use in half but double your acreage, the Ogallala gains nothing. If you triple your acreage, the Ogallala has actually lost. Drip irrigation is very close to 100 percent efficient, but it brings marginal lands into production using marginal wells. Without drip available, that water would have stayed in the ground.

"They've talked an awful lot around here about getting more conservation-minded and saving water," sighs Dean Yonts. "But I think we finally found some individuals who understood that if we lined the canals and saved 50 percent of the water, we didn't really save anything. We simply changed the timing of when that water would get downstream. That's what the center pivots are doing. Rather than putting twenty-four inches of water on a field and have the crop use twelve, like we used to with furrow, the pivot operator will put on fifteen inches in order to let the plants use twelve. The rest of it, that added amount between the twenty-four and the fifteen, is still upstream. Eventually, it's going to have to make its way downstream. You'll end up with the dams in Wyoming going over the spills a lot more, and bringing that water directly down the river."

"Or you might have the other nine inches consumed locally, by somebody who isn't as efficient an operator," adds Steve Sibray. "There's a lot of unmet demand out there."

Yonts nods. "So how much water do you save?"

"Are you saying that saving water wastes water?" I ask.

Sibray smiles. "You bet. By increasing your efficiency, you can actually increase your consumptive use. It's like the steam engine. The first steam engines were very inefficient, so the consumption of coal was minimal. James Watt invented a very efficient steam engine, and consumptive use of coal greatly increased. Technology has an impact, but it's not always the obvious one."

XIX

MAKING EVERY BIT COUNT

I F WATER USE CANNOT be reduced by deploying more efficient equipment, how is reduction to be achieved? Ask a water professional today, and you are likely to get a one-word answer: management. Any old irrigation technology—even the much-maligned flood system—can deliver water to the crop in an efficient manner. It is the choices you make while operating the system that count.

In Colorado, Joel Schneecloth reminisced about a project undertaken early in his career. "It was surface irrigation, but we were doing top management on everything," he told me. "Ten miles away, a farmer had installed drip on a quarter-section. The first year he had it in there, he said, 'Man, the water savings are tremendous! I only put sixteen inches of water on with my drip compared to thirty-two with my furrow, and I got two hundred bushels of corn per acre!' I had to laugh, because with our flood irrigation—we had designed it right, and managed it right—we had put twelve inches of water on the ground, and we had made two hundred and ten bushels per acre. Of course, there's that difference of ten miles, so you don't know exactly what it was. But I think it all comes back to management."

"We can and we do overirrigate with our furrow irrigation systems," admits Scottsbluff's Dean Yonts. "But other than maybe that first or second irrigation of the year, when we're really into putting the water on, the rest of the time we're probably not overirrigating to the extent that you might think. Conversion from furrow to sprinklers, saving water, not saving water—the whole issue is really what plant you are growing, and how much water that plant consumes, because that plant is the only real user of that water."

So if the key is going to be good management, what does good management look like? We can theorize; we can speculate; or we can go to a region where groundwater decline has already made good management necessary and look around.

AT THE CORNER OF 19TH STREET and Avenue G in downtown Lubbock stands an eight-foot-tall statue of Buddy Holly. Exhibits in the nearby Buddy Holly Center inform you that the release of the rock 'n' roll icon's 1957 megahit *That'll Be the Day* was a defining moment in the history of popular music, and that the guitar techniques, vocal methods, and performance style of Lubbock's proudest product were pivotal influences on a couple of Liverpudlians named John Lennon and Paul McCartney who (the exhibits say) chose to call their group the Beatles as a tribute to Holly's backup band, the Crickets.

The statue is wearing horn-rimmed glasses and holding a Fender Stratocaster guitar.

The Brownfield Highway begins a few blocks away, at the corner of 4th Street and Boston Avenue on the north side of the Texas Tech University campus. Slicing diagonally across Lubbock's neat north-south street grid, the highway leaves the city at its bottom left-hand corner and arrows southwest across the Llano Estacado. The fields begin at the edge of the city and sweep from the shoulders of the road to the far horizon, as flat as the face of Holly's Stratocaster. Almost all of them bear a single crop: cotton.

We think of cotton as a crop of the hot and humid South, so it may surprise you to find that the center of American cotton production, for most of the past half-century, has been the southern High Plains. Texas is our number-one cotton-growing state by a wide margin, harvesting over a billion dollars' worth of the fluffy white fiber every year. There are counties in Texas that outproduce entire southern states. Cotton was King in the antebellum South because the Peculiar Institution of slavery gave planters a ready supply of cheap labor, and because the temperature range was right for the plants. The humidity range, however, was all wrong. Most parts of Dixie drown in at least fifty inches of rain each year. Cotton does best between twenty-five and thirty, and it will produce a crop on as little as thirteen.

Lubbock gets a shade under nineteen inches of rain per year, midway between what cotton needs and what cotton wants. Most of that rain obligingly falls during the growing season, which runs from the beginning of May to the middle of October. It does not hurt that nearly all of it comes in the form of "intense rainfall events"— thunderstorms—with copious sunshine the rest of the time to ripen the crop rapidly and prevent boll rot. Given these climate conditions, it should not seem strange to find that, of the fifteen top cotton-producing counties in Texas, thirteen are blocked solidly around Lubbock. In 1997, cotton farmers in Lubbock County alone harvested 280,000 acres. The county's next largest acreage that year was sorghum, at 24,000.

"It's a cotton desert down there," observes a water official in Amarillo. "You've got the stock-farming mix up here; you go south, you go to the cotton desert."

The official is speaking metaphorically, of a land where nothing significant grows except cotton, but there have been many years when he could have been speaking meteorologically as well. Like all areas that depend on episodic rainfall, Lubbock's annual precipitation varies enormously. The city sees bumper years, but it also sees drought years—years when rainfall is less than 75 percent of normal. Under light drought conditions, cotton watered by rain alone will barely produce; under moderate-to-heavy drought, it will die. To make it through the dry years, farmers in the cotton desert have come to depend on another source of water: the Ogallala Aquifer.

Thus have the fortunes of King Cotton become thoroughly entwined with the fortunes of declining groundwater. It is a gradual, ongoing ensnarement. Farmers put in irrigation systems as insurance against dry years, and then—because they have them, because they desire a return on their investment, because they can—they use them in wet years as well. Cotton does all right on Lubbock's average rain-fall, but it does much better on a little bit more.

IF LUBBOCK IS THE HEART of King Cotton, the Brownfield Highway is his aorta. In the thirty-eight miles from Lubbock to Brownfield—where it is called Lubbock Road—the highway cuts diagonally through three of the thirteen cotton-desert counties. The other ten either border or corner this central trio. When I drove the highway one blazingly hot September morning not long ago, the impression I came away with was not so much a cotton desert as it was a cotton sea. The horizon formed a perfect circle; the cotton, heavy with the coming harvest, faded to blue vagueness in all directions. Breezes moved through it, causing chop, like wind over deep water. Birds circled as if looking for islands.

There were just two jarring notes. One was the high-voltage line that paralled the highway, its skeletal towers marching in lockstep into the distance. The other was the center-pivot sprinklers. They were idle, that season, and they loomed in the fields like vast, abandoned insect husks, enormous but dead. Come April, I knew, they would lurch to life once more, but the lives they lived would look somewhat different from the lives of sprinklers farther north.

"What I work for is to try to help the farmers adopt conservation strategies so they'll be able to get by with less water," explains Dana Porter. A tall young woman with ash-blond hair and a Ph.D. in agricultural and biological engineering from Mississippi State University, Porter is the current occupant of Leon New's old position at the Lubbock Extension Center. Growing up on a farm near Amarillo, she herded cattle and learned to swim in tailwater pits ("that was before chemicals were bad for you"). She has worked in both New Mexico (dryer than Texas) and West Virginia (much wetter); now she is back in the panhandle, trying to help farms like the one she was raised on stay viable.

"In the southern part of the aquifer we're already practicing deficit irrigation," she told me. "That is the rule, not the exception. Up north, where they have more water, it's kind of a novel, neat new idea to have managed deficit irrigation, where you only provide a certain percentage of the crop's demand. That's been the rule here for a while, because it's all the water we've had."

There is an optimal amount of water, called crop water demand, that will lead to optimal yield; both irrigating more and irrigating less will decrease the harvest. Crop water demand varies with a host of factors, the most important of which are crop species and climate. Alfalfa needs more water than corn; corn needs more water than cotton. And alfalfa, corn, and cotton all need more water on a hot, dry day than they do on a cold, foggy one.

Different crops also vary in their sensitivity to receiving less-than-optimal water, a quality known as drought tolerance. Research conducted in southwest Kansas in the late 1990s suggests that corn is moderately to highly drought tolerant: the researchers were able to maintain 90 percent optimal yield on 50 percent of corn's crop water demand. Alfalfa, by contrast, is drought intolerant: yield falls off faster than water reduction, with a 10 percent loss of yield occurring after an 8 percent reduction in water. Most crops fall somewhere between these extremes. Soy and wheat are good examples: with 75 percent of their crop water demand—a figure cited by Porter as a reasonable target for deficit irrigators in the Texas Panhandle—each of these common High Plains crops will produce 90 percent of its optimal yield. Total harvest drops a bit, but the efficiency through which water is converted to crop volume is significantly improved.

Cotton, like corn, is moderate to highly drought tolerant; it is able to produce good yields on as little as half its crop water demand. The key to success is timing. Cotton will root well and produce stalks and leaves on very little water. When irrigation is really needed is while the plants are in flower: Seven inches of water applied in the three weeks after the first bud (or "square") appears, but before the plant reaches full flower, will do more for crop yield than any amount applied before or after.

Most crop plants are similar to cotton in that their water requirements increase and decrease at specific times in their growth cycles. A large part of good irrigation management, therefore, is determining how much water a crop actually needs *right now*. Modern irrigation farmers depend heavily on computer software, which combines data on the type of crop; the date of sprouting; the length of the local growing season; the temperature, field capacity, and soil moisture tension; and the amounts and timings of any rainfall and irrigation to date, and spit out precisely when and how much irrigation is going to be needed.

"When I start with a deficit amount, I have to make every bit of it count," Porter explains. "If I plant a crop but I can't irrigate it all season, then I've wasted seed, and I've wasted the water I put on in the early part of the year. So what we're trying to do is figure out how many acres a farmer can manage adequately through the season. Irrigation is one of the most expensive inputs that we have, so we had better make it count. If we had plenty of water, we might not be so sensitive to that."

ALTHOUGH FARMERS USE computers, they may not have the training to use them properly, and they usually don't have the time to gather the data. That job is normally done by a crop consultant, an independent contractor with the equipment and expertise needed to do the job right. The farmer makes the final decision, but it is the crop consultant's data, and often the crop consultant's computer printout, that the farmer works with.

"You get a field map," explains attorney and irrigation farmer Mike Ramsey in Garden City. "You get soil moisture content in two or three places in the field. The consultants monitor the top three feet of soil and tell you how much reserve moisture you have for the plants. They also give you the evapotranspiration rate, so you know what you're going to use for the next week, assuming normal weather conditions."

A difficulty with this last is that weather conditions on the High Plains are seldom normal: They are usually too hot or too cold or too dry or too damp. Rain, in particular, is a problem. Ogallala irrigators cannot afford to follow the old advice to "leave room for rain"; too often, the rain doesn't come. When it does, it is sometimes heavy enough to be a disaster. Crop consultants can tell you how much water your soil and your crop need. They may be able to predict how much rain you will shortly receive, and whether that rain will be helpful or dire. What they cannot do is to turn it on and off.

"Part of the problem of managing a center-pivot system," complains Ramsey, "is rain events. You're trying to maintain an optimum soil moisture level for the plant, for agronomic reasons. You have this sprinkler that's going in a circular motion at pretty much a steady rate of speed. Right behind the sprinkler you've got a refilled profile, and five feet in front of it you've got a deficit just sufficient to refill without overfilling. Along comes a rain event, and what happens? You wet the whole field, and you bring the leading edge up to saturation. But your trailing edge was already at saturation, so you've overwatered that. And then you have to wait enough time to restart the system so that the leading edge will take water. You try to time that restart so you're not overirrigating at the beginning, hoping by the time you get all the way around again you won't have a deficit. It can be very tricky. It would actually be better if it never rained, from a center-pivot irrigator's point of view. In the perfect world, you'd have an irrigation system that could put the water on instantaneously, and then turn off. Not only put it on instantaneously, but put it through the soil profile in all the right places without any losses to evaporation. But the technology can only go so far, and then Mother Nature and management get in the way."

Ultimately, that is what is going to determine the future of the Ogallala Aquifer: Mother Nature and management. Technology is impartial: It can help, but it can also hinder. Proper management is necessary to make certain that water-saving devices actually save water. But even proper management cannot help if the rains come at the wrong time, or come too hard, or fail to come at all.

XX

A TENUOUS BALANCE

I T IS IMPORTANT TO NOTE that when someone says the Ogallala is running out of water, it is not the same thing as saying there
will be no more water available from the aquifer at all, forever
after. "Running out of water" really means running out of *stored* water.
The aquifer will continue to act as a conduit: Whatever flows into it
will always be available to flow out again.

Before irrigation pumping began, the water table was holding
steady, with discharge through springs exactly matching recharge
through rainfall. There are signs that equilibrium could return again.
The water table is much lower now and still declining, but in many
areas the decline appears to be leveling out. We may be approaching
what can be termed a "developmental equilibrium": an equilibrium in
which the water table again holds steady, deeper in the earth—much
deeper—but no longer dropping.

Several things may be contributing to this new, deeper steady state.
These include (in no particular order):

- *Reduced spring flow.* With less of the recharge leaving the system
  through springs, more is available for maintaining the water
  table. Because the springs had value, for plant life, wildlife, and

stream flow, this is not necessarily a desirable thing; but it is there, and it may be having a beneficial effect on Ogallala water levels.

- *Return flow from irrigation.* Not all water pumped from the Ogallala into a center pivot is lost; some of it trickles down through the soil and returns to the aquifer. This downward per-colation can take many years, so its effect was not immediately noticeable when large-scale pumping began in the 1960s. But return flow has now reached the water table almost everywhere, and that is undoubtedly helping to slow the decline.

- *Developmental saturation.* In the sixties and seventies, new wells were being drilled at headlong speed all over the aquifer, and each one meant additional pumping and an additional cone of depression. That frenzy has now passed. There are simply not very many places for new wells to go these days. In some areas, they are forbidden by law; in others, they are legally allowed but economically unattractive. Existing cones of depression have stabilized, and few new ones are being added. With the excep-tion of places such as the Nebraska Sand Hills, which still have an essentially virgin Ogallala supply, development is pretty much over.

The fact that decline seems to be stabilizing, though, does not auto-matically mean that everything will be fine on the High Plains ever afterward. Questions must still be asked. If a new equilibrium is achieved, will it be at an acceptable level? If the level is not acceptable, can we find a way to raise it? How long can we expect it to last before drawdown resumes?

The answers to all of these questions are troubling. The equilib-rium we appear to be approaching is acceptable only if the empty lake behind Optima Dam is acceptable, only if dry springs and negligible stream flow are acceptable, only if massive expenditures of fossil fuel to pull water from deep in the earth to maintain an ever-more precarious food supply are acceptable. Raising the water table has proved notoriously difficult, and the few places where it has worked—usually by accident—have had quality problems. And the balance, if it exists, is a tenuous one. The leveling of the rate of decline promises to be extremely short-lived. The problem is no longer limited to the old one of agricultural overuse. That has not gone away, but it has been joined by others. Few of these are limited, as agriculture is, to flat, tillable land during the growing season.

There is, for example, bottled water. Americans love it. We spend more than seven billion dollars per year to drink water from the supermarket shelf instead of the tap. Bottled-water consumption in the United States in 2002 was 21.2 gallons per capita, up nearly 11 percent from the year before. There is no reason why wellhead owners on the High Plains should not take advantage of this lucrative new market; some have already done so. There are plants bottling Ogallala water in both Texas and Nebraska. These are small operations, but they could grow. A following could develop for Ogallala water, similar to the following for Kona coffee or for Napa Valley wine. With water in twelve-ounce bottles currently selling for three times the price of gasoline, there is every incentive for High Plains well owners to encourage this development.

The Texas portion of the aquifer may be particularly at risk because of the Rule of Capture. One of the defining pieces of case law governing that rule, in fact, concerns bottled water. Since 1996, a company called Ozarka, a subsidiary of international water giant Perrier, has been pumping a large volume of water from a small piece of land in a Dallas suburb. Shortly after pumping began, Ozarka was sued by a group of its neighbors for drying up their wells. The case ended up in the Texas Supreme Court, which ruled in Ozarka's favor in 1999; the Rule of Capture, the justices complained, allowed them no other choice. In Texas, as Ozarka's Lauren Cargill put it bluntly after the verdict, "You can do with groundwater what you want, regardless of your neighbors." Ogallala water was not involved in this particular case, but you can be certain that Ogallala well owners were listening.

Bulk shipment of Ogallala water—transporting it by tanker instead of in bottles—has also been seriously proposed. It almost achieved reality in 2002. That summer, as Denver and other cities along the Front Range suffered through a record-setting drought, a Sedalia, Colorado, businessman named Robert Krumberger formed a company called Homeland Hydro Options to explore methods of bringing water to the stricken region. The scheme he eventually settled on was to drill three wells in the heart of the Sand Hills near Hyannis, Nebraska. The water from these wells was to be pumped into railroad tank cars and transported to Denver. Each car in this "rolling pipeline" would hold 30,000 gallons; there would be one 96-car train each week. Roughly nine acre feet—2.9 million gallons—would leave the aquifer every time the train ran. Krumberger faded in the face of fierce opposition from Nebraskans; his wells were never drilled. That does not mean that the idea has gone away.

A third nontraditional use for Ogallala water is river augmentation. States are often obligated by interstate compacts, or by the terms of legal settlements, to deliver specific volumes of water at specific points in a river's course—usually at state lines. Rather than shut down withdrawals of river water upstream, it may be easier to pour groundwater into the river at the point where the higher flow is required. That is a concern in New Mexico, where Ogallala water has been proposed to be used, like steroids, to bulk up the Pecos River just above the Texas state line (see chapter 8). It could also happen on the Canadian, to settle complaints from environmentalists that irrigation withdrawals are threatening habitat for an endangered fish called the Arkansas River shiner; or on the Arkansas, to satisfy the terms of the legal settlement Kansas recently won against Colorado; or on the South Platte, to allow

Denver to dry up the river upstream but still allow water to flow into Nebraska.

Or it could happen on the Republican. That would be particularly interesting, because it is groundwater withdrawal that has caused the Republican's problems in the first place—and groundwater augmentation that is contributing, contentiously, to its tenuous survival.

XXI

AN ELEPHANT IN A DARK ROOM

O NCE WHEN JIM GOEKE was out of town I spent most of a rainy Wednesday in North Platte attempting to find someone else—anyone else!—who would admit to knowing something about groundwater issues in Nebraska. I began at the University of Nebraska Research and Extension Center on State Farm Road, where Goeke has his office. The agronomist and the soil scientist I spoke with there both suggested that my best course would be to wait for Goeke to come back. I asked if anyone knew anything about the related matter of the Republican River lawsuit. They said that would be Goeke, too; he was on the scientific panel that had been set up to help settle it. I went to the North Platte library. The clerk at the circulation desk directed me to the Research and Extension Center and suggested that the best person to talk to would be a geologist named Goeke. Driving through downtown North Platte earlier, I had noticed a Nature Conservancy office. I called the office, hoping to speak with someone about the effects of groundwater decline on natural areas. "That would be Jim Goeke," said the helpful voice on the other end of the line. "He's on our board of directors."

When Goeke returned that evening, he asked how the day had gone. "Not one of my better days," I replied. "Everywhere I went, everyone I asked about the Ogallala kept referring me to some guy named Jim Goeke."

Goeke laughed ruefully. "It's a curse," he said. "A curse for me and a curse for the people who keep leaning on me." I didn't have the heart to remind him that the curse extended well beyond North Platte. A few days earlier I had been in Holdrege, on the plateau between the Platte and Republican rivers in south-central Nebraska. The region around Holdrege is one of the few parts of the High Plains irrigated by surface water—in this case, water from the North Platte River, caught behind Kingsley Dam north of the city of Ogallala and transported south and east through more than 150 miles of canals. I went to the headquarters of the Central Nebraska Public Power and Irrigation District, known throughout Nebraska simply as Central, which holds responsibility for the canal system. Central's youthful information officer, Jeff Beuttner, was willing but apologetic. "If you're talking to Jim Goeke, I probably can't help you very much," he said. "He knows more about our system than we do."

Beuttner's statement may not be literally true, but Goeke certainly knows a great deal about the effect of Central's system on the Ogallala Aquifer and how that, in turn, has affected the Republican River lawsuit, a case to which both the history and the future of the Ogallala are inextricably tied.

THE REPUBLICAN RIVER'S NAME does not originate from the political party but from the Kitkehahki band of the Pawnee Indians, whom early European settlers—having read their Plato, and believing they had stumbled across an Athens in the wilderness—insisted on referring to as the "Republican Pawnee." It is a tristate stream. All of its three principal branches originate in Colorado: the North Fork in Yuma County, the South Fork and the middle fork (which for some inexplicable reason is called, not the Middle Fork, but the Arikaree River) just a few miles apart in Lincoln County, on the western cusp of the High Plains. The Republican proper is said to begin where the Arikaree and the North Fork come together, just inside Nebraska; the South Fork joins twenty miles downstream, near the town of Benkelman. A bit over four hundred miles below that, at Junction City, Kansas, the Republican combines with the Smoky Hill River to form the Kansas River. If you were to walk along the Republican's bank from its mouth at Junction City to its most distant source at the headwaters of the Arikaree, you would have trudged a total of 568 miles.

Two additional bits of trivia seem worth mentioning. The first is that the Republican's basin contains the lowest piece of land in Colorado: The bed of the Arikaree at the point where the river leaves the state, in a little canyon called the Arikaree Breaks twelve miles southeast of Wray. The second is that the river was the locale of Kansas's only gold rush. During the Great Depression, rumors spread of color in the Republican's sands, and droves of hopeful Kansans with gold pans descended on the waters of the South Fork. University of Kansas geologist E. D. Kinney joined them briefly, dipping a pan himself to test the rumors. His report: no gold, "desireable as this would be." The value of the Republican, like other High Plains streams, would lie not in its bed, but in its water. And there is enough value there to keep a large phalanx of lawyers thoroughly occupied for a long, long time.

"THE LEGAL ASPECTS OF WATER," points out Goeke, "are driven by hydrological extremes." For that reason alone, one would expect extensive legal snarls to entangle the Republican: Its hydrology is among the most extreme of any river in America. During normal times, you can wade across the little river at just about any point along its length. During its flood of record, the main stem carried the highest stream flow ever recorded in Nebraska—280,000 cubic feet per second, roughly eight times the average pre-dam flow of the Missouri River. That figure was recorded at the end of May 1935, when a damp storm moving westward from the Great Lakes collided with an equally damp storm moving northward from the Gulf of Mexico. The divide between the North Fork and the Arikaree took most of the fallout. Official rain-gauge records for that period do not exist, but an unofficial gauge set out by a Colorado rancher collected 24 inches of precipitation in 24 hours. Eyewitness accounts describe a wall of water shaped like a spearhead advancing down the valley of the Republican with a roar that could be heard from several miles away. At the town of Cambridge, roughly halfway through the Nebraska section of the river's course, the flood's rapidly moving front was two miles wide. More than one hundred people—and more than twenty thousand head of livestock—lost their lives.

That 1935 flood "just traumatized the hell out of people," says Goeke today. "It was probably a five-hundred-year flood. Wall to wall. And then in 1947 there was another one. Fourteen or fifteen inches of rain fell in the headwaters of Medicine Creek, and a tidal wave of water came out of the hills and down into the Republican Valley at Cambridge. It was a nice sunny day, and a wall of water hit that town and killed fifty-some people. That set the foundation for a lot that went on afterward."

THE FEDERAL GOVERNMENT has always claimed authority over navigation on the waterways of the United States. Historians of water law are fond of pointing out that one of the first acts of the nascent U.S. Congress in 1789 was to pass a law authorizing spending for seaway navigation, in the form of a lighthouse on Chesapeake Bay. The murkier question of whether federal authority should extend inland, to navigable rivers and lakes as well as to the ocean, was firmly settled in the affirmative by the Supreme Court in 1820. During the middle of the twentieth century, federal authority was extended to water pollution and to endangered aquatic species. All remaining issues regarding water have traditionally been left to the states. Federal agencies usually build flood control and hydropower projects, but these are always—in theory, at least—approved first at the state level.

Leaving water issues to the states works adequately when the water involved is within the boundaries of a single state. Matters become more complicated when the water flows from one state into another. In such cases, the states are not equal: What a downstream state does to a river has few effects upstream, but what an upstream state does has a great many effects downstream. There exists a hand-wringing letter written in 1887 by Nebraska governor John Thayer to Colorado governor Alva Adams, in which Thayer pled with his fellow governor "in the name of humanity" to order Colorado irrigators to let some of the river flow into Nebraska. Adams gave the letter to his attorney general, Alvin Marsh, who replied to Thayer that, while he sympathized with the plight of Nebraska irrigators, his hands were effectively tied as long as Colorado irrigators were acting in accordance with the law. This state of affairs—the wringing of hands downstream and the tying of them upstream—continues to prevail today. Although states can sue one another, and have done so with some frequency, these suits rarely have the effect the downstream states desire.

What Marsh wrote to Thayer has largely been upheld by the U.S. Supreme Court, which sits as the trial court for legal actions brought by one state against another. The court's earliest controlling opinion regarding interstate water disputes was issued in 1907, in regard to a suit brought by Kansas against Colorado over the Arkansas River. Equal access to the water of a river, the justices ruled, didn't mean the states had to divide the water equally; it only meant that the water-rights laws of one state couldn't be more favorable to private individuals than were the water-rights laws of other states on the same river. Faced with this reasoning, states largely gave up on the federal government and began instead to negotiate compacts.

A compact is essentially a treaty—a legally binding agreement among two or more governments. Treaties between constitutent states of the Union are expressly forbidden by Article I of the Constitution, but there is a loophole: One state may enter into "an Agreement or Compact" with another if Congress consents directly to each agreement. So all interstate compacts have Congress as a silent partner. Compacts actually exist as federal laws; states pass enabling legislation to put the federally enacted compacts into practice, rather than passing the compacts themselves. Typically, compact compliance is monitored by a commission whose members are appointed by the governors of the states involved, with at least one federally appointed member sitting in as well.

THE REPUBLICAN RIVER COMPACT was signed in 1943. By that time, several active irrigation canals laced the river's valley and several more were set to be built. Two—one existing, the other proposed— were of particular interest to the compact's negotiators, because they crossed state lines. The existing canal, the Pioneer Ditch, originated on the North Fork near Wray, Colorado, but carried most of its water into Nebraska, where it was used to irrigate lands near the junction of the North Fork and the Arikaree. It was operating in the shadow of litigation that had been unresolved since 1922. The other, the Courtland Canal, was still just a gleam in its promoters' eyes. When completed, in 1950, it would divert water from the river at Guide Rock, Nebraska, to irrigate about six thousand acres in Nebraska and more than eleven times that much in Kansas.

The compact apportioned the river among the three states of its basin according to a complicated series of hydrologic formulae. The Pioneer Ditch and the Courtland Canal were given specific allocations. Flows were then designated for the main stem and for each major tributary, based on the river's "virgin water supply"—the water supply "undepleted by the activities of man"—during an average year. In what turned out to be a key provision, Article III allowed adjustment of the allocations if the river's flow changed:

Should the future computed virgin water supply of any source vary more than ten (10) percent from the virgin water supply as hereinabove set forth, the allocations hereinafter made from such source shall be increased or decreased in the relative proportions that the future computed virgin water supply of such source bears to the computed virgin water supply used herein.

There was also a key omission. The compact didn't mention groundwater at all. At the time, it hardly seemed to matter. Ten years later, along came Frank Zybach and his sprinkler.

THE REPUBLICAN RIVER COMPACT stipulated a "virgin water supply" of 478,900 acre-feet per year at the Kansas–Nebraska state line. By 1959, it had become apparent that the river was rarely going to provide that amount. That triggered the provisions of Article III, and a tristate commission, the Republican River Compact Administration, was put in place to allocate the river's water. It was only partly successful. Sprinklers were proliferating in southwest Nebraska, and Kansas, convinced that its upstream neighbor was siphoning water away through the compact's groundwater loophole, was becoming angrier and more obdurate each year. And then, in 1972, a Chase County, Nebraska, farmer named Joy Sporhase put in one more sprinkler, and all legal hell proceeded to break loose. It was not the amount Sporhase was pumping from beneath the land he co-owned with his son-in-law, Delbert Moss, that created the uproar. It was the location of the sprinkler's pivot—precisely fifty-five feet from the Colorado state line.

Sporhase and Moss had purchased their land at auction, then found that it lay in two states. That created problems when they decided to irrigate. Phillips County, Colorado, was closed to new high-capacity wells. Nebraska allowed them, but the Chase County portion of their land was too small to make irrigation worthwhile. The logical solution was to irrigate the whole thing from a well on the

Chase County side. And that put Sporhase and Moss in conflict with Nebraska law every time their center pivot went around.

Nebraska's water code contained what is known as a reciprocity clause: Water could be transferred out of the state if and only if it could also be transferred back in. Sporhase and Moss were transferring water out of the state precisely because it *couldn't* be transferred back in. Colorado wouldn't allow them to sink a well on the west side of the state boundary, so they couldn't put a pivot there and rotate it into Nebraska. The reciprocity clause was violated. Nebraska sued to shut the operation down.

"The Nebraska Supreme Court upheld the idea of nonreciprocity," Goeke explains. "Sporhase and Moss appealed, and it went to the U.S. Supreme Court. And when it came back from the Supreme Court, the lower courts' decisions had been overturned, on the basis that groundwater is an article of interstate commerce, and you can't restrict the use of it, or its movement across state lines, unless the health and human welfare of the people at the point of origin are endangered." The court's ruling, which came in 1982, stated that the reciprocity clause was "an explicit barrier to commerce" and was therefore in violation of the commerce clause of the U.S. Constitution.

"There was a lot more to it," adds Goeke, "but all of it dealt with the transfer of water across state borders. It said that political boundaries could not be a constraint to the movement of water. Water's a transient resource—it doesn't recognize state boundaries." The legal requirement to treat water as transient holds whether the water involved is groundwater or surface water, *and* whether or not there happens to be an interstate compact in place that governs the second but says nothing at all about the first.

IN FEBRUARY 1997, fifteen years after the Sporhase-Moss ruling, Kansas finally abandoned the Republican River Compact Administration. David Pope, the state's Republican River adminstrator, wrote a letter to the other members of the panel explaining his government's position:

> Resolution of Kansas' concerns regarding enforcement of the Republican River Compact requires Nebraska not only to recognize and understand the problem, but to take meaningful action towards its resolution. Despite the hard work and the good intentions of the Nebraska negotiating team, we believe that, until the responsible parties of the State of Nebraska acknowledge their obligations under the Republican River Compact, and take action to fulfill them, our continued participation in negotiations will not lead to agreement or action. As a result, Kansas will no longer participate in the mediation process. Kansas will continue to put its energies into exploring other means for resolving our concerns.

There was little doubt about what "other means" meant. "When Kansas walked away from the process," remarks Goeke, "it was obvious that they were considering filing suit."

The suit was brought by Kansas attorney general Carla Stovall a year later. In her brief, Stovall accused Nebraska of consistently using more water than it was entitled to under the Republican River Compact. There were two suggested mechanisms, both related to groundwater pumping.

The first mechanism involved wells in the alluvial terraces beside the river. These wells utilize aquifers that are hydrologically connected to the stream flow. The Republican is a losing stream through most of its length, so the water in the alluvial aquifers was actually coming from the river, and Kansas wanted it counted toward the total that Nebraska was allowed to withdraw under the compact.

The second mechanism involved wells in the uplands. These, the brief stated, were dropping the water table in the Ogallala and drying up the headwater springs of the Republican's tributaries, which—as gaining streams—now had a smaller reservoir to gain from. Water that should have been flowing into the tributaries, into the river, and eventually (here was the kicker) into the Courtland Canal at Guide Rock was flowing out of center-pivot sprinklers in Nebraska instead.

TO ADJUDICATE INTERSTATE disputes, the Supreme Court employs a person called a "special master"—someone from outside the region of the dispute who is qualified, by training and experience, to weigh the legal issues involved and to offer recommendations to the court. In theory, the court may ignore these recommendations; in practice, it seldom does, so the choice of special master is crucial. For the Republican River suit, the court chose Vincent L. McKusick of Portland, Maine. McKusick's legal credentials were impeccable: a former chief justice of the Maine Supreme Court, he had earlier served as a law clerk to two eminent figures in American judicial history, Judge Learned Hand of the U.S. Court of Appeals and Justice Felix Frankfurter of the U.S. Supreme Court. He also had a master's degree in engineering and a background in public utility law, and was likely to understand the science behind the dispute better than most justices could.

McKusick issued his initial ruling on February 12, 2001. Neither state was happy. Nebraska had been contending that groundwater was not included in the compact; the special master said it was. That meant that the compact's water budget—the balance sheet, similar to a money budget, that keeps track of where all the water in a river basin comes from and to whom it is allocated—would have to be recalculated to take into account springs, seeps, and wells along with stream flow, evaporation, and direct irrigation from the river.

That was what Kansas got. It also got a ruling that Nebraska's total water withdrawal from the Republican Basin—surface water plus groundwater—could be no more than the amount allocated in the compact. What Nebraska got, in turn, was recognition that the original total streamflow calculation—the "original virgin water," as it is usually called by the compact partners—still held. Those figures could not be recomputed, as Kansas wished, to allow for groundwater withdrawals. A hydrologic study would be needed to find out if, when, and by how much the original virgin water amounts had actually been violated by Nebraska's wells.

"Before Kansas walked away, there had been some talk of the three states getting together to do a basinwide study," notes Goeke. "And the politics were such that we were now discouraged from being involved in that. If the end result played poorly for Nebraska, tax dollars might be seen to have been used to undermine Nebraska's position. So the United States Geologic Survey, as an objective third party, was brought in to do a basinwide study." He waits for the complications this implies to sink in, and then adds, with a chuckle, "If you want a yes or no answer at some point today, you let me know."

THERE WERE TWO additional complications in the Republican River lawsuit—untidy snarls of geography and politics that the special master's preliminary ruling had not addressed. The first of these concerned the Central Nebraska Public Power and Irrigation District's project at Holdrege. The second concerned the cryptic but inescapable fact that there were actually *four* states in the basin of the Republican River—and that Kansas was two of them.

The Holdrege situation is complex, but it can be summed up in a fairly straightforward manner. Holdrege irrigators receive water from the North Platte River and discharge it onto their fields. Some of that North Platte water ends up in the Republican River.

If you examine a map depicting changes in groundwater levels in the Ogallala Aquifer, at almost every place you look you will see declines. There are a few spots, however, where you will actually see increases—places where the water table is rising instead of falling, forming a hill of groundwater that bulges above the prevailing regional water table like a big underground bubble. One of the largest of these bubbles, which geologists call "groundwater mounds," lies beneath Holdrege. Its boundaries correspond with remarkable precision to the boundaries of the irrigated lands that get their water from the Central Nebraska Public Power and Irrigation District.

"Yeah, that's our water," Central's Jeff Beuttner agreed cheerfully when I asked him about it. He went on to explain its origin, which was about what you might expect. The Holdrege-area canals date from an era when no one was particularly concerned about lining irrigation ditches to prevent leakage. The unlined canals have been leaking ever since, and the leaked water has been trickling down to the Ogallala and building up the mound. The mound drains to the south, boosting stream flow in many of the tributaries that feed the Republican. Beuttner remarked that Central had filed a friend-of-the-court brief in the litigation over the Republican River Compact: The agency thought that as long as groundwater effects were being brought into the picture, Nebraska should get credit for the water that Central was putting *into* the Republican as well as the water that others in the state were taking out.

In North Platte a few days later—as Beuttner had predicted he would—Jim Goeke explained the Holdrege situation in considerably more detail.

"When we signed the compact, we didn't know much of anything about the available water supply in the Republican Basin," Goeke began. "Our records only go back to the early 1930s. We signed a compact that committed us to apportioning the flows of the river based on barely ten years of flow records, and with little or no record of groundwater resources at all. And the ink was hardly dry on that compact when they started to deliver water from the North Platte to Holdrege. Kingsley Dam was put in, water was diverted down a canal along the south side of the South Platte, brought over into Sutherland Reservoir, and taken down into central Nebraska. And in central Nebraska, the canals that delivered the water started to leak. The water went down and built that mound, and the mound drained south, into the Republican. And the compact talked about the 'virgin water supply unaffected by the activities of man.' That's all we were committed to."

Water leaking from Central's canals has had a significant effect on the Republican's flow. "As this mound has built," Goeke notes, "the streams coming out of that upland area have actually grown. Their headwaters have moved to the north, and their flow has increased considerably. And one of the bones of contention is that Nebraska ought to get credit for that extra flow—the *non*-virgin water supply that *has* been affected by activities of man. It ought to be a credit to Nebraska, because it moves from the Platte to the Republican, and adds to the flow of the river."

In Holdrege, Jeff Beuttner told me that Central was trying to protect and conserve its groundwater mound, lining ditches in areas of high groundwater and leaving them unlined in depleted areas, managing the recharge that was feeding the mound as much as they were able. Part of that management required negotiating with Holdrege-area farmers who, wanting the labor-saving advantages of center pivots, had been pulling themselves off Central's flood system and sinking wells into the mound, reducing recharge (less water running through the leaking canals) and increasing discharge (more water being pumped out of the mound). To combat this, Central had developed a method that allowed sprinklers to run off ditch water. They were doing everything they could, Beuttner insisted, to keep those Republican River tributaries flowing with Platte River water, to keep the interbasin transfer alive, to keep the Republican alive. As we spoke, the sun shone on Holdrege, a polished, picture-perfect High Plains town. There was a large camp for German prisoners of war just outside Holdrege during World War II, and the town and its surroundings were so pretty that, following their official repatriation to Germany after the war, many of the former POWs came right back here and settled down. It is far too attractive to be the bone of contention it has since become in the growing dispute over the distribution of Ogallala groundwater.

COLBY, KANSAS, SPREAD contentedly over the low loess hills that top the Ogallala south of the Republican Valley, also seems far too pretty to be a bone of contention. Like Holdrege, however, it has become one. The blame for this lies less with the town's activities than it does with its geography. Colby is the economic center of the Republican Basin's "fourth state"—the portion of Kansas that lies not at the river's mouth but near its headwaters. The South Fork and the Arikaree both pass through the northwest corner of the Sunflower State on their way from Colorado to Nebraska, picking up several tributaries along the way. Because of this, a small but significant part of the reduction in the Republican's flow into Kansas is due, not to irrigation in Nebraska or in Colorado, but to irrigation in Kansas itself.

"Development in northwestern Kansas has dried up the tributaries on the south side of the Republican," contends Goeke. "That has contributed to the diminished water supply in the river. So Kansas, in fact, has dirty hands. They're part of their own problem."

In Colby, the Northwest Kansas Groundwater Management District's Wayne Bossert does not disagree with that. "Kansas is both an upstream and a downstream state, because the Republican comes in and out and back in," he explains, quietly. "The original complaint came from the eastern part of Kansas, and that was mainly a response to what the Nebraska portion was doing. But you know, you can't divorce from the fact that we're above Nebraska."

Bossert is slumped behind his desk in the groundwater management district's office, sitting almost on the back of his neck, his fingertips pressed professorially together. "To mitigate that fact, the state engineer has—rather artfully, I think—moved some stuff around. It was advantageous to all three of the states to do some . . ." A pause. "I don't want to call it 'creative accounting.' It's kind of creating accounting, but it isn't like Enron creative accounting. It's flexible accounting that still meets the essence of the goal. All three states have an advantage that way. We can trade back and forth, as long as the whole system works. That has helped our area. But we have to be part of the solution for northeast Kansas." He brightens a little. "Our numbers aren't really that bad. And I have great hope that, in the average to wet year, we're not going to see much impact. But when there are two, three, or four dry years together, the entire system is going to feel it. And if you're going to summarize all of that in generic language, I believe the chief engineer has mitigated us the best that he possibly could, but we're not absolved in all situations."

VINCENT McKUSICK SET March 1, 2003, as the date to begin the court proceedings in the Republican River dispute. It didn't happen. In late October 2001, the three states entered negotiations to settle out of court. Six weeks later, they asked the special master to suspend the suit pending the results of the negotiations; a year after that, on December 16, 2002, they announced a settlement. The key to the settlement was a computer model of groundwater flow that was satisfactory to all of the parties. The catch was that the model hadn't been completed yet.

The model was still in flux when I talked about it with Goeke four months later. "The whole thing is going to revolve around a model agreeable to all three states," he explained. "All the parameters that go into that model have to be acceptable to all of the parties involved, and most of them are only estimates. We can measure stream flow—that's something tangible that we can get with a degree of confidence. But the distribution of rainfall, the amount of recharge . . ." He shook his head. "Recharge is a hell of a kicker. And how much water is actually being pumped? There's another one, because a lot of it isn't metered. Those are real problems, as far as reconciling the different states' takes on it."

The connection between surface water and groundwater is another factor on which it can be difficult to reach agreement. The geology of a riverbed may remain the same for many miles, or it may change dramatically within a few feet. That alters the amount of water the river exchanges with the aquifers it passes over. "People have a tendency to look at a river and think rivers are all the same," sighs Goeke. "But when you consider the Platte, and what the Platte crosses—the landforms, the runoff, the whole damn thing—it's just an amazing stream in its relationship to the saturated thicknesses that it runs over. Whereas the Republican, in many places, basically has nothing. The Republican is underlain by Pierre shale, and by the Niobrara Formation, which underlies the Pierre, and there is no connection between the river and the groundwater. There's alluvium—river-deposited soil—that extends through there, so you've got some continuity. But pumping in the alluvium has no connection with the upland surface, the plateau above the valley, whatsoever."

"So there's not much of a tie between the Ogallala to the south of the Republican and the Ogallala to the north," I venture.

Goeke leans back and puts his hands behind his head. "There's none. There is no Ogallala in that valley. We've drilled the shi—" He catches himself. "We've drilled a lot in that valley, and there isn't any Ogallala down there. You drill through alluvial soils, you drill through sixty or seventy feet of river gravels, and boom, you drill right into the shale. And a lot of people are willing to disregard the geology, because who cares? So much water in, so much water out. But the transfer and timing of the water really depend on the hydraulic properties of the materials that you've got. And we probably have more information about that than Kansas does, or Colorado.

"You know that story about the elephant in the dark room? The man who feels the elephant's trunk thinks the animal is like a hose, the one who feels its side thinks it's like a wall, and the one who feels its tusk thinks it's like a rock, and so on? I use that analogy all the time on the Ogallala. Because somebody down in Texas sees a whole different portion of the elephant than what we see, or what Kansas sees, or what Colorado sees. The reality in northwestern Kansas is that they've got disconnected streams—their streams have ceased to flow. In Nebraska, our streams are still supported largely by groundwater discharge. Down there the drainage is from the southwest to the northeast, and they've got dry streambeds. Here we have live streambeds, and instead of draining from southwest to northeast they drain from northwest to southeast. It's a whole different reality."

Every outcrop, every test well, every meter on the pad of a center pivot is a grope in the dark toward a different part of the Ogallala's anatomy. Is the aquifer thick or thin, sand or gravel, healthy or scraping bottom? It depends on which part of the elephant you happen to have hold of. In Texas, Ray Brady once showed me three wells ten feet apart, all of them seven inches in diameter, all of them drilled into the Ogallala Aquifer. The water level in one well was eighty feet higher than the levels in the other two.

"They're drilled to different depths," he explained. "You've got a red clay layer on top, and the two deeper wells take water pretty far below that. The shallow well goes just five feet below the red clay layer. The water levels in these three wells all started off at the same point. First year of pumping, two of them lost forty-five feet."

"Forty-five feet!" I marveled.

"Yeah. First year," said Brady. He smiled. "People are always asking me how deep it is to water in the Ogallala, and I ask them, 'Where are you standing?'"

Texas's wells are not the only ones that show this somewhat bizarre behavior. In Kansas, I was told of a well near Stafford whose cone of depression, it had been feared, might be drawing down the water level of its immediate neighbor. They test-pumped it for five days. There were no effects on the well next door, but a well a mile away dropped two feet. In Nebraska, I heard of a man who had sold a house with a rock-solid-dependable well in its basement. He went half a mile away to build another house. When he began to drill his new well, the rock-solid-dependable well at the old place went dry within fifteen minutes. Groundwater does not move that fast, but pressure changes in a confined aquifer can and do.

Those are the parts of the elephant that the designers of the model for the Republican settlement are up against. The modelers are using datum points one mile apart. That is a vast improvement over the ten-mile grid used for the High Plains RASA. For a river basin with a total area of 22,400 square miles, it is certainly practical. But it cannot begin to account for the complexity of a hydrologic system in which the water levels in two wells ten feet apart can differ by eighty feet.

THE SIGNIFICANCE OF the Republican River settlement extends far beyond the confines of the river's basin. There is now precedence in interstate law for recognizing that groundwater pumping in one state can affect the water supply of another, and that states may be held liable for the damage this causes to their neighbors. That could have a profound effect on developments along the Texas–New Mexico border. The settlement establishes, as a matter of law, that water table decline can have an adverse affect on surface waters, which may shift the outcome of cases revolving around wildlife habitat loss due to declining spring flow, or around reservoir depletion in regions of heavy groundwater irrigation. For these things, we can probably thank Vincent McKusick's engineering background. Science and law should never be adversaries. When they quarrel—as they often do over the Ogallala—lawyers are the only winners.

As part of the settlement, Nebraska has declared a moratorium on new wells in the Republican Basin and has agreed to install meters on all existing wells. Metering will supply accurate data for the modelers, but it will also allow pumping to be regulated—not a popular notion. "They'll probably come up with an allocation," Goeke observes. "People don't like it. But for years they've had free access to the water, and the freedom to waste it if they wanted. And I don't know if Big Brother's a particularly good system. But there are things afoot that are going to impose efficiencies of use. If you're not wasting water, you're saving money. I think profitability and economics are going to drive a lot of this."

Will economics really cause people to waste less water? Is profitability a blessing to the Ogallala, or a curse? These questions are now being tested. One of the principal testing grounds lies well south of the Republican River, in Ray Brady's part of the Texas Panhandle, where a dynamic has developed that involves money, water, politics, and something that would be described, in any other context but that of the Rule of Capture, as wholesale theft.

XXII

ROBERTS COUNTY RANCHER

W HO OWNS THE OGALLALA? As we become more and more dependent on its output and less and less certain of its future, this question takes on increasing urgency. Should underground water be a public resource, as it is in six of the eight High Plains states, or should it belong to the owner of the overlying earth, as in Oklahoma, or to no one, as in Texas? If domestic or livestock wells dry up in the vicinity of an irrigated field, is the owner of the irrigated field liable for damages? All states require permits for high-capacity wells; should the permit process be extended to include the well in your backyard, or your pasture, or a state park campground? All states similarly require any water pumped from the ground to meet the test of "beneficial use." Which uses are beneficial? A free-flowing stream? A golf course? A fishing lake? When the water needs of two beneficial uses conflict, is one use more beneficial than the other? Who makes that decision?

As we struggle with these issues, it is necessary to keep the larger context in mind. The question that must be dealt with is not whether or not any specific use of the Ogallala Aquifer will survive. It is not even whether or not the aquifer itself will survive. The question is whether or not the systems dependent upon the aquifer—the

natural systems, the agricultural systems, and the human systems—
will survive in a form resembling their current state. Or in any form
at all.

That is why the question of groundwater ownership is so important on the High Plains. In a dry land, water is coveted. If ownership is
not clearly established, it is easy to take the water away.

BOONE PICKENS WOULD PREFER that you not think of him as a
corporate raider. "I never liked being called a raider," he has complained. "I never destroyed anything."

"Dallas oilman" doesn't please him much either. "Geologist" is a little better; he has a degree in geology, and he retains his membership in
the American Association of Petroleum Geologists. But the days when
he had much passion for field work are long past. His self-image, these
days, lies elsewhere.

"You know how I should be identified?" he says firmly. "Roberts
County rancher."

I note that he has owned a ranch in Roberts County since 1971.

"That's right," he says. "But they don't want to put that in the
stories—it legitimizes me too much. They would rather have me as an
oil baron. But really, when you get down to it, I'm a Roberts County
landowner."

It is some piece of land. Sitting astride the Canadian River in the
upper right-hand portion of the Texas Panhandle, Roberts is one of
more than one hundred little square counties in west Texas that were
partitioned off in the late 1870s and named for mostly minor heroes
of Texas history. Because no attention was paid to the land when the
borders were drawn, these counties suffer from topographical quirki-
ness. Roberts is among the quirkiest. In a region famous for flatness,
Roberts County has almost no flat land. Most of it is occupied by a
slanting amazement of gullies and gorges and ridges that twist down
to the Canadian River, eight hundred feet below the Llano Estacado.

Pickens's ranch lies near the heart of that hummocked terrain, its south end in the sky, its north end dipping into the river. The county covers 924 square miles. Pickens owns thirty-eight of them. Beneath those thirty-eight square miles lies a thick, nearly untouched portion of the Ogallala Aquifer.

Four hundred river miles below Roberts County, the small town of Holdenville, Oklahoma, huddles on the low crest of the long divide between the Canadian River and its principal tributary, the North Canadian. Holdenville is where Boone Pickens learned the value of a dollar. His father, Boone Sr., was a small-town lawyer and a dabbler in the oil bidness; his mother worked for Franklin Roosevelt's Office of Price Administration. Boone Jr. carried newspapers and mowed lawns. There still wasn't much money. Eventually, the family moved to Amarillo, where Boone Sr. thought prospects might be brighter. Only five-foot-nine, skinny but pugnacious, the new kid in town went out for the high school basketball team. Coach T. G. Hull put him in at guard. It is probably no coincidence that the team made it to the semi-finals of the state tournament that year. Boone Pickens has never liked to lose.

After an abortive year at West Texas State College in Canyon, twenty miles south of Amarillo, Pickens entered the geology program at Oklahoma A&M College (now Oklahoma State University) in Stillwater. Following graduation (with honors) in 1951, he took a job with Phillips Petroleum. It was not a match made in heaven. The oil giant was rigidly bureaucratic—there seems to have been more paperwork than field work—and the $290 per month that Pickens was paid as a junior geologist wasn't much of an inducement to stay. He took three years of that and quit. Two years later, on the strength of twenty-five hundred dollars in borrowed funds and support from a couple of businessmen, one of them his wife's uncle, he started a company called Petroleum Exploration, Inc. In 1962, its sixth complete year of operation, P.E.I. earned more than three-quarters of a million dollars.

NINETEEN SIXTY-TWO was also the year Ray Brady graduated from high school. This was in Crystal City, Texas, a little cattle town on the Nueces River eighty miles or so north of Laredo. He was an indifferent student. I once commented to him that my wife and I had earned our diplomas at about the same time as he did. "I don't know if y'all had as much fun as I did," he drawled, "but I did graduate." Following graduation, he worked as a heavy-equipment operator, a school-bus driver, a cowboy, a janitor; he took livestock classes at Uvalde Junior College. After a couple of years of kicking around that way, he enrolled at Texas Tech in Lubbock to study agricultural engineering.

In 1966, in the middle of his second year at Tech, Brady found himself drafted. He applied to the Army Corps of Engineers' Officer Candidate School and surprised himself by being accepted. The army gave him a year of training and sent him to Vietnam as a combat engineer with the rank of second lieutenant. He cleared mines and built water-treatment plants. Back stateside but still in the army, he returned to Texas Tech under an armed services program called Operation Bootstrap. "I had a choice," he says today. "I could go back to Vietnam, or I could go to school." A smile. "I thought about that for a long time."

To qualify for Operation Bootstrap, you had to pick from a slender list of academic fields. Agricultural engineering wasn't on it, so Brady chose geology. "I took freshman and junior geology courses one year, took sophomore and senior courses the next go-round—took four years of geology in two years, met the requirements, and got the degree." The army made him a captain and sent him off to Korea to probe for tunnels under the DMZ.

BOONE PICKENS, MEANWHILE—untroubled by the draft—was busy making money.

That's not quite the way he would put it. He saw his role as making money for the stockholders in his company, not for himself. Most companies, he believes, do a lousy job of that. "Chief executives," he once famously remarked, "who themselves own few shares of their companies, have no more feeling for the average stockholder than they do for baboons in Africa." He was determined to do right by his. In 1964, the year P.E.I., now renamed Mesa Petroleum, became a publicly traded company, its gross revenue was $1.5 million. Eight years later, it had an annual gross of nearly $100 million and a net worth almost twice that amount.

Much of that success can be attributed to the same scrappy pugnacity that helped propel Amarillo High School to the state semifinals. Boone Pickens did not invent the art of the hostile takeover, but he certainly lifted it to new levels. He saw what he was doing as a crusade on behalf of the stockholders of poorly run companies, and he went at it with a kind of cocky zeal that, he admits today, was shamelessly arrogant. "Back then, nobody could tell me anything," he remarked not long ago to *Fortune* Magazine's Joseph Nocera. "I knew everything. I didn't, of course. But I thought I did." That attitude would bring him multiple millions, legendary status in American business history, and a *Time* magazine cover graced with his face. It would also bring him a messy divorce and a brutal comeuppance engineered by a former underling named David Batchelder, who used the techniques he had learned at the master's feet to force him out of his own company.

By 1988, the year the disillusioned Batchelder left Pickens's employment, the ride was essentially over. Pickens had failed in six straight takeover attempts. The first four had at least enriched Mesa stockholders; the next-to-last barely broke even, and the last one lost big. The company began hemorrhaging money. A very public feud with the *Globe-News* in Amarillo, which had been home to Mesa since its inception, caused Pickens to pick up his headquarters and move them expensively to Dallas. His second marriage broke down; he underwent treatment for depression. In 1996, Batchelder saw his chance and made his move. Pickens was dislodged from his position as Mesa's CEO. A short time later, he resigned from the board and left the company altogether.

Back in 1971, as part of a move to diversify Mesa into cattle operations, Pickens had purchased twenty-eight thousand acres of the Canadian River Breaks in Roberts County. The diversification failed— the cattle lost money—but the land took hold. Pickens kept the property, naming it the 2-B Ranch: one B for Boone, the other for Bea, his second wife, whom he was then in the process of marrying. The 2-B ran a few cattle, but its main product was solace. It became a retreat, a place where he could watch the wildlife, play a little golf—he installed a two-hole course—and generally forget the corporate world. When his company and his marriage both fell apart, he sold the four-thousand-acre piece of the ranch that contained the house he had shared with Bea, built a new, larger home on the twenty-four thousand acres that remained, renamed the spread Mesa Vista, and began spending most of his time there.

THE YEAR THAT DAVID BATCHELDER left Mesa, 1988, was also the year that Major Raymond Brady left the U.S. Army. He had given it twenty-two years of his life, and it was time to do something else.

The first thing he did was to go back to school. Enrolling once more at Texas Tech, he completed a master of science in civil engineering in 1991. "I took full advantage of the GI bill," he says today. "I said, 'I'm gonna use *all* of this.' And then I came up here and worked at the neighborhood nuclear weapons plant for five years." "Here" is Amarillo, and what Brady calls the "neighborhood nuclear weapons plant" is Pantex, a sprawling, sixteen-thousand-acre facility northeast of the Amarillo airport that is America's only atomic-bomb factory. Brady, with his unusual combination of geology and engineering skills, was put in charge of groundwater monitoring. "Didn't really want to do that," he observes, laconically. The part of the Ogallala Aquifer beneath Pantex was so heavily contaminated that, in the third year of Brady's employment there, it was declared a Superfund site. The water contained trichloroethylene, chromium, lead, and methylene chloride. Plutonium was stored in concrete bunkers dating from World War II. Playa lakes—the main recharge mechanism for the Ogallala on the Llano Estacado—were being used as wastewater sumps. "Ninety-two was a wet year," Brady recalls. "A lot of the playas filled up. We had four of them, and they all went back to normal at different rates. The one that emptied the fastest was the one they were discharging the treated wastewater into." The implications of that for Brady's job were disconcerting. Discharging wastes regularly into the playa, he con-cluded, was keeping the soil column saturated, which accelerated the speed at which the wastes moved into the soil. Pantex had super-charged its pollution. Making his rounds one blisteringly hot Texas day, Brady noticed a jackrabbit resting in the shade of a telephone pole. Each time he passed the pole, the shade had shifted a little, and so had the jackrabbit. Working at Pantex had begun to feel that way—constantly shifting position to stay out of the heat. He resigned and went back, once more, to Texas Tech. This time the degree he came out with was a master of engineering. His thesis dealt with the effects of metal corrosion on groundwater monitoring wells.

WHILE BOONE PICKENS was retiring to Roberts County to lick his wounds and Ray Brady was trying to figure out how a fun-loving south Texas ranch boy had managed to find himself monitoring groundwater pollution at America's only nuclear weapons plant, the city of Amarillo was slowly running out of water.

That's an oversimplification, but it is reasonably close to the truth. The general alarm over plunging water levels in the Ogallala during the sixties and seventies had passed, leaving a High Plains population warily watching its water as the wells inched further down. Most of the decline was felt by agriculture, but municipal water systems were not immune. This was especially true in Texas, where groundwater is governed by the Rule of Capture, also known as the Law of the Biggest Pump. If your neighbors' wells are drawing down the water table on your property in the Lone Star State, you cannot sue them; all you can do is try to pull it out through your well faster than they are pulling it out through theirs.

There is in Amarillo a restaurant called the Big Texan Steak Ranch, where they will give you a seventy-two-ounce steak for free if you can eat the whole thing (along with a baked potato, a salad, and a shrimp cocktail) inside of an hour. That should give you some idea how important cattle are to this region, which raises roughly 25 percent of all beef consumed in the United States. It has been said, with some truth, that there are more cows than people in the Texas Panhandle. Cows drink more water than humans do. They also eat irrigated crops, and vast amounts of water are used as they are processed into steaks and ground beef and spare ribs. Before you bite into your next cheeseburger, consider: That sandwich—bun, cheese, tomato, lettuce, mayonnaise, and all—cost approximately seven hundred gallons of water to produce. The meat patty alone was responsible for more than six hundred of those.

Cotton and peanuts are also grown in the panhandle, along with a small amount of corn and milo. Cotton and milo can be grown here without irrigation, but they usually aren't; peanuts and corn always require extra water to survive. The extra water comes from the Ogallala Aquifer. And because of the Rule of Capture, all of those agricultural superstructures—those crops, those cattle, those packing plants—are in direct, aggressive competition with municipal well fields.

Amarillo is not fully dependent on well fields: The city also has a share of Lake Meredith, an impoundment of the Canadian River a few miles north of town. Lake Meredith water is controlled by an entity called the Canadian River Municipal Water Authority, abbreviated as

CRMWA, which everyone pronounces "crumwa." Amarillo is one of eleven member cities of CRMWA, which also serves Plainview, Lubbock, and several points farther south, so what little river water there is has to be spread pretty thin. For that reason, Amarillo also has its own water agency, Amarillo Water. Amarillo Water gets all of its contribution to Amarillo's water supply from the Ogallala Aquifer.

By the early 1990s, it had become apparent to both utilities that they needed help. CRMWA was especially vulnerable. The area had just been through one of the wettest winters in its history—1991–92, the same season Ray Brady was studying Pantex's playas—but Lake Meredith had resolutely refused to fill all the way. CRMWA's hydrologists had begun to suspect that it would *never* fill all the way. It also had a quality problem. Water pumped from it was saltier than federal health standards allowed, courtesy of a saline aquifer in New Mexico. CRMWA was pumping water out of the saline aquifer, which had slowed its discharge somewhat, but that was an expensive and inadequate solution that left them with a growing sea of brine that was eventually going to get into the river anyway. What they really needed was cleaner, fresher water to mix with Lake Meredith water to lower its salinity. And another Amarillo-area utility, it turned out, just happened to be sitting on a huge amount of the stuff.

In the early 1970s, Southwestern Public Service (SPS), the electric utility that serves the city of Amarillo, had begun planning a nuclear power plant. The operation of such plants requires a great deal of water, mostly to cool the active uranium in the core. Because of this, they are normally built near large lakes or rivers. None were available in the Texas panhandle, so SPS planned to obtain its water from the same place most people in the region did: the Ogallala Aquifer. The utility put together a wholly owned subsidiary, Quixx Corporation, to purchase the necessary pumping rights.

Quixx operatives quickly realized that water was going to be cheaper

where there were no competing uses for it. On top of the Llano Estacado near Amarillo, irrigated farmland and beef feedlots were taking
nearly every available drop, the city was taking the rest, and the Ogallala was in decline. On the steep, broken terrain of the Canadian River
Breaks, irrigation had proved next to impossible, the population was
small and widely scattered, and the aquifer was largely intact. Quixx
wheeled, and it dealed, and it eventually came up with drilling rights to
110,000 acres of the breaks. All of them were in Roberts County.

The nuclear plant was never built. Like people in other parts of the
country, Texans had grown increasingly suspicious of nuclear power;
SPS took the prudent course and scrapped its plans before construction began. Quixx was put on standby. The drilling rights, which had
already been paid for, sat idle. Twenty years later, those idle rights
came to CRMWA's attention.

In 1993, the water company and the electric-company subsidiary
put together a deal. Stripped of legal minutae, what the deal did was to
give CRMWA nearly forty-three thousand acres' worth of the drilling
rights held by Quixx, and to give Quixx fourteen and a half million
dollars. CRMWA authorities announced that they would start mixing
Roberts County water with Lake Meredith water as soon as they could
drill the necessary wells and construct the necessary pipelines.

No groundwater management district covered Roberts County in
1993, but there was one next door: the Panhandle GMD, the third oldest in the state, dating back to 1955. When news of the pending
CRMWA–Quixx deal leaked out, a group of Roberts County residents
quickly mounted a drive to annex themselves to the district, and on
May 7, 1994, Roberts County voted to join the Panhandle GMD and
come under its management rules. One of those rules prohibited the
transport of groundwater pumped from any well in the district to any
location beyond the district's boundaries. CRMWA's system would
be transporting it as far south as Lamesa, more than three hundred
miles away.

CRMWA and Quixx campaigned vigorously against the annexa-
tion, but they also pursued other options. On April 15, 1994, three
weeks before the election, they quietly filed suit against the Panhandle GMD in district court in Amarillo. The suit claimed immunity
from Panhandle rules, on the grounds that the drilling rights being
transferred to CRMWA had been purchased by Quixx long before the
election. In case that argument failed, the company's brief also con-
tested the validity of the rule forbidding out-of-district transport.
That fallback argument did the trick. When the suit finally came to
trial in late 1995, Judge Patrick Pirtle ruled that the Panhandle GMD
had the right to regulate CRMWA's use of Quixx's water. But he also
stated that

> any rule attempting to regulate or prevent the transportation of
> water out of the district is *ultra vires* [beyond its power] and
> invalid. . . . The district does not have the authority to regulate or
> prevent, by permit or otherwise, the transportation of water out-
> side the district.

Judge Pirtle's ruling had two immediate effects. The first was that
CRMWA was able to conclude its deal with Quixx and begin drilling
wells and building pipelines. The second was that Boone Pickens
woke up.

THE PROPERTY TO WHICH CRMWA had obtained drilling rights
lay just three miles southeast of Pickens's Mesa Vista Ranch. Pickens
is an intelligent man with a degree in geology, and he fully understands groundwater dynamics. Once pumping began, he quickly realized, the declining water levels in CRMWA's wells would cause water
to flow toward them. Toward CRMWA's wells was away from the
Mesa Vista. Water would be drained out from under the ranch, fun-
neled into CRMWA's pumps, and sent south to Lamesa. It was theft,

but the Rule of Capture legalized it, and there was nothing Pickens could do about it.

If there is anything that raises Pickens's ire faster than abuse of stockholders' rights, it is abuse of property rights. He wasn't using the water under his land, but dammit, it was *his*. Why should he lose it without a dime in return? And brooding about that, Pickens realized there was something he could do after all.

He could sell the water before CRMWA drained it away.

"I think they awakened me to the opportunity," says Pickens today. "I don't think I was smart enough to figure it out by myself, even though I knew the water was there. It would never have occurred to me to put together a deal to transport water to south Texas. I was ignorant of water needs over the state; I'd just never looked into it. But once I was being drained, then I was really stupid if I *didn't* look into it, and start to find if there was some way I could sell water to offset the drainage. And that's the way it unfolded."

He went first to CRMWA, offering to sell the right to pump directly instead of waiting for the water to drain sideways three miles. CRMWA wasn't interested; the Quixx deal plus the Rule of Capture had given them all the water they thought they would need. His next try was at Amarillo Water, which had just announced that it would be purchasing the water rights on the narrow strip of land between the CRMWA property and the Mesa Vista. Amarillo turned him down, too. So Pickens took the one step still open to him.

He formed a company to market the water himself.

The company, Mesa Water, was incorporated in late 1999. It was considerably more than a one-man operation: Realizing early on that water rights on just twenty-four thousand acres weren't enough to attract the kind of large buyers he had in mind, Pickens had approached his neighbors. Eight of them joined Mesa Water, granting Pickens effective control over water rights on roughly 150,000 acres. One of the coalition's new members was Quixx; in a masterstroke worthy of his corporate-takeover days, Pickens had managed to co-opt the remaining assets of CRMWA's principal supplier.

Mesa Water moved quickly to establish its presence. On September 18, 2000, a company representative went to the Panhandle GMD's headquarters in the little town of White Deer and dropped off nine separate permit applications for high-capacity wells, on behalf of Pickens and the other eight members of what has since come to be called the Mesa Group. The permits were accepted, stamped, and taken down a short hallway to the desk of the GMD's recently-hired assistant manager and chief geohydrologist: Major Raymond Brady, USACE (ret.).

"WE'RE A LITTLE EARLY FOR FLOWERS," says Brady. "We've had freezing temperatures this week, so we're just a little bit early. Monday and Tuesday, temperatures were in the high twenties out here. Week before last, highs during the day were in the eighties. So it's spring in the panhandle."

We are somewhere in the breaks above the Canadian River, in a north-facing maze of gullies and hills under an Ogallala blue sky etched by delicate white runes of cirrus. Brady points to a fence. "The south side of that fence is still Quixx," he remarks, "and it's totally tied up with Mesa. The north side is CRMWA. From here on north, every-thing's in a deal of some kind. See that green stake out there with a white deal on it? That's one of our rain gauges. We've got those scattered all over the district, but I've put in a few extras around here, because of the interest in this business."

The ground is animal-colored, a thousand shades of brown, black, and tan, with occasional patches of green in the draws. It fades to purple in the distance, where we can see the flat, even top of the North Plains. The Ogallala Formation dips low here, running beneath the river, connecting north and south, North Plains and Llano Estacado, with sinews of groundwater.

"It's the same formation from here all the way to Nebraska," notes Brady, "but probably with different ages." He pauses. "You know, there's always the question: if the water level goes down in Nebraska, is it going to affect Roberts County? The answer is no. It's just too far. How fast does groundwater move? Hundreds of feet a year. So if you go down fifty feet in Nebraska, even overnight, what's that gonna do down here? Nothing. It's just too far. And you've got discontinuities in surface elevation as well as water table elevations between here and there, so you'd have to overcome all of those. It just doesn't happen. See the bird?"

"What kind was it?" asks my wife.

"Cheep-cheep bird," says Brady. He grins at her questioning look. "I don't know what they are. Little birds that go cheep-cheep. Here's another one of our monitoring wells. CRMWA was required to drill five of them as part of their permit conditions. We watch these wells pretty closely, because they're going to tell us what the effects of the project are."

We pass a rancher pitchforking hay out of the back of a pickup. "These guys watch their grass," Brady comments. "Of course, basically, they're harvesting grass."

"Through cattle," I observe.

"You got it. It's not overgrazed. What happened in the twenties and thirties I don't know, but the land's got good turf now. And if you've got good turf, then if you graze, you tend to hold the water a little bit longer; you don't get as much washing away downhill. Always a delicate balance."

As we approach the river, the land seems paradoxically to be growing dryer. The predominant plants are sagebrush and yucca, widely spaced. Beneath them, on the surface, are the pinkish Ogallala sands. "This ranch right here just sold," Brady observes, nodding to the right. "The fellow who had the surface apparently bought it after the drilling rights were sold, and he sold it last year to another fellow. So the surface has changed hands at least twice since the drilling rights have."

Soon he notes that we are passing over land belonging to Warren Chisum. Chisum, a Republican, represents the fifteen northernmost panhandle counties in the state legislature down in Austin. I recall a conversation with Boone Pickens a month earlier. We had been talking about media coverage of Mesa Water, and I mentioned an article I had just seen in the *Austin American-Statesman* that quoted some of his opponents. "Yeah," said Pickens, "in fact, they even quoted my state representative."

"Yes, I think they did."

"Warren Chisum." Pickens said the name as if he were spitting out a grape seed. "They didn't say this in the article," he added after a moment, "probably because the reporter didn't know to ask the question. But Warren Chisum doesn't have any water. He sold it to Amarillo. And he doesn't want me to have the same opportunity he had."

"Yeah," agrees Brady now, "this went to Amarillo. Depth to water here is about eighty feet. You saw another monitoring well back there? This year—first year of operation of CRMWA's project—it's gone down four feet."

"That would seem to validate Pickens's concerns."

"Well, yes." Brady keeps his eyes on the road. "I think everybody is concerned. We know some of the things that have happened in the past, but we've never really had a chance to watch one of these closely from the get-go. That's why I've got this thing surrounded. And we've got wells within the project, too. Not as many as I'd like, but we do have some. So the question then becomes, At what point does the influence spread outward?" He gestures to the right. "Over here there's 421 acres right in the heart of all these water deals. The owner will not sell. That is some of the deepest, thickest water out here. Depending on whose numbers you believe, he's sitting on anywhere from four to eight hundred feet of saturated sand."

"What's he keeping it for?"

"Doesn't think it's right to sell it."

"He's providing an undisturbed island," observes my wife. "Which may erode."

"Well, yeah. It'll erode." There is no doubt whatsoever in Brady's voice. "He very graciously allows us to monitor some of his wells. Truth be known, we could probably require him to do that, but that's not been our policy. There's another of our monitoring wells, on top of that hill." He gestures toward a bluff on the left. A small white object clings to its northern lip. "That one's pretty far north, but it's strategically located, because this is the end of the Amarillo property. We're about to cross over to Bill Tolbert's, which is in Mesa."

The road swings to the right and comes down to a small bridge. Not the Canadian, yet—Chicken Creek. There is a copse of large cottonwoods just upstream protecting a patch of lush, damp-looking grass. Brady comments that there were once sawmills along the Canadian that specialized in cottonwoods, and that now few trees like these remain. The road swings left again, onto the right-hand slope of the valley, still trending north and down. A large wooden sign appears beside the road in the middle distance, next to a green gate.

"Okay," says Brady. "Now you're looking towards Mr. Pickens's stuff. That's his gate, down there. When they laid out the section lines here, they laid out river sections. Half a mile wide, two miles long. Of course, a lot of the surveys were done in *varas*. You know the *vara*? Spanish measurement. Thirty-three and one-third inches. Remember the Republic of Texas? Before the Republic of Texas, Mexico, and before that, Spain. Here's the gate. There's his mesa and everything. There's his vista of his mesa."

The gate, on the west side of the road, is built of welded two-inch pipe hinged to a metal post probably nine inches through. Its olive-green paint is fresh. Close behind it looms the sign, rising on massive wood columns out of a pile of quarried boulders. Big letters proclaim

MESA VISTA RANCH; smaller letters add ROBERTS CO. TEXAS. PRI-
VATE ROAD. NO TRESPASSING. Behind the sign the land makes its final
descent to the river, which winds unseen beneath tall purple bluffs
perhaps five miles away.

Brady is talking about the Mesa Vista. "Pickens has developed his
ranch as a quail plantation, for lack of a better word," he says, looking
out over the brown and green land between us and the river. "He's got
a pipeline network out there—takes a backhoe, makes a little scoop,
and puts in a drip to make pools for his quail. Interesting operation."

I recall how, in our conversation, Pickens had referred to quail as
his livestock. "And really, that's true," agrees Brady. "In a lot of this
country. Where I grew up, in south Texas, the deer hunting business is
now better than the cow business. My father and I sold cattle in the
early sixties for sixty-five dollars and up. Last year, cattle were still selling for sixty-five dollars and up. That gives you an idea of how profitable the cow business is. Not very."

The economics Brady cites may have something to do with Pickens's decision to run quail instead of cattle. The fact is, though, that
the man genuinely likes quail. He allows much of the ranch to be used
by Oklahoma State University as a quail study area. "The primary use
of my water is for my wildlife," he once told me. "Other people are
more interested in livestock than they are in wildlife. That's fine—
whatever suits you best. But what I do for my wildlife is comparable to
what you'd do for your livestock." There are wetlands in the bottoms
of some of the draws on the Mesa Vista property. I asked Pickens if he
thought they would survive Mesa's pumping. "I do," he said. "And I
don't think that Amarillo or CRMWA are going to drain that groundwater reservoir to where I couldn't have water for wildlife, either."

Whether that statement is true or not may depend to a large extent
on Ray Brady and the Panhandle GMD. The Rule of Capture prevents
groundwater management districts from regulating how much water
may be drawn from any individual well, but it does not prevent them

from regulating how much may be drawn from beneath the district as a whole. The Panhandle district has such a regulation, put in place in 1998 and called the 50/50 Rule. It is more complicated than I am about to state it, but the gist is that the saturated thickness of the part of the Ogallala Aquifer that lies within the district may not be reduced by more than 50 percent before 2048, fifty years following the rule's adoption (50 percent in fifty years: hence, 50/50). Practically speaking, what that means is that well owners are only allowed to pump water from the top half of the aquifer. If the 50/50 Rule holds, the Mesa Vista wetlands will probably survive. If it does not— if saturated thickness declines below 50 percent—the future becomes more doubtful.

Some have questioned the validity of the 50/50 Rule, citing the Ogallala's extreme variability. "What the hell does that mean?" asked an incredulous Jim Goeke when I brought the subject up to him. "Fifty percent saturated thickness? That doesn't mean diddly. What happens if the upper fifty's all gravel, and the bottom fifty's a silty sand? You're screwed, because taking that upper fifty means you've taken the entire future out of the aquifer. And if the upper fifty's a fine-grained silty sandstone and the bottom fifty is gravel, you've left a gold mine on the bottom and taken the crap off the top. Saturated thickness doesn't mean anything unless you consider the character."

Brady doesn't disagree with this assessment. "Mr. Goeke is absolutely right," he wrote to me some time later:

> The tradeoff is getting something into the rules via the political process. The details of "enforcing" are (in my opinion) a combination of geology, engineering, legal opinions, voodoo and bluff. We will get our chance sooner or later to try all this. It will be fun. My opinion is, it is better to try something than to continue to ignore the problem & hope it goes away. The 50/50 rule is what we were able to get away with.

Pickens, speaking for a moment as the geologist he once was, agrees at least partially with Goeke. "Fifty percent of saturated thickness," he told me, "is going to be very, very hard to determine." Nevertheless, he supports the rule. "We have no problem with 50 percent. We believe that's a good rule. I think the ideal situation would be that everyone who wants to sell water would all have the same allowable volume to pump, just as you would in an oil field. You would produce your water, and they would produce their water, and you'd draw down to 50 percent and then stop producing it. Fifty years from now, things may be changed to where recharge would be meaningful to the reservoir, and there could be another round of selling water." He chuckled. "I realize that in fifty years I won't be here, so I won't have to deal with that problem."

Pickens is not so much concerned whether or not Mesa Water will have to abide by the 50/50 Rule as he is whether or not CRMWA and Amarillo will have to. "All we asked the groundwater district is, 'Look, just give us the same deal you gave the other people.' That's all we asked. We didn't want anything extra. But they didn't give us as good a permit as they gave CRMWA or Amarillo, because those outfits get to produce 100 percent of their water."

"What you're saying," I said, "is that CRMWA isn't bound by the 50 percent rule."

"That's correct," said Pickens. "Nor is Amarillo. Of course, nobody ever says *they're* gonna drain the aquifer, they say *I'm* gonna drain it. It's bizzare. The way it's described, I'm gonna turn this area into a dust bowl. They aren't even gonna have enough water to drink. Just unbelievable statements made." He reined himself in with difficulty. "Anyway, I expect the district to do just what they said. They said they are not going to take more than 50 percent out of the reservoir. But CRMWA and Amarillo, neither one will agree to that."

Standing with Brady next to the Mesa Vista's gate, I bring up Pickens's concern. Brady looks thoughtful. "Here's the story," he says after a minute. "CRMWA's claim is that they got a permit before the rule was in place, therefore they're exempt. That's one way to look at it. But if you read their permit, it says 'subject to the continuing rules of the district.' I don't know if it will eventually be settled in court or not. But our interpretation is that they're subject to the continuing rules of the district, and the 50/50 Rule is a continuing rule."

"How about Amarillo?"

"Amarillo's permits came after 1998, so they're definitely covered. Of course, you've got to remember that Amarillo is a member city of CRMWA. There's two pheasant hens, by the way." Two large brown birds whir by, necks outstretched, headed for Chicken Creek.

"So you're going to try to hold Amarillo and CRMWA to the 50/50?"

Brady nods. "We intend to enforce the 50/50 Rule. If we don't we might as well just shut our operation down and go in the house. Because if you don't A. have a conservation goal; B. have a conservation rule to enforce that goal; and C. have the will to enforce it, then you're wasting the taxpayers' money. We intend to enforce it. I think that CRMWA understands that we intend to enforce it. But that leaves the question, What are we going to do? We'll see. Right now lawyers are all making lots of money, analyzing and sending briefs and comments and questions back and forth. Whether the 50/50 rule is the best policy philosophically I don't know, but that's been the tradition since this whole thing started."

We get back in Brady's pickup and head toward the Canadian. "There are two areas across the river," Brady continues, after a bit. "One's thirty-thousand-some acres and one's twenty-thousand-some. About fifty-five thousand acres in all. They're owned by Harold Courson, and he has a permit to develop them. Of course, they, like Quixx, are quote-unquote 'independent' of Mesa, but I have to remember that the day they brought the applications in, they were all delivered by the same person in the same envelope. I don't know what that means, but I believe I can figure it out."

"Actually," I remark, "Pickens was pretty clear about that. Courson's with him."

Brady nods. "I don't know the details, and really, I guess, it's none of my business. But for whatever reason, they're making sure that everybody has an individual permit."

"It's probably something—"

"It has to do with lawyers in expensive shirts and shoes and ties."

All of us laugh. "Lyndon Johnson," I observe, "is once supposed to have said that if a town in Texas was too small to support a lawyer, it could always support two."

Brady chuckles. "Yeah. Interesting folks. Lawyers are people who send you faxes at 5:01 on Friday afternoon."

LAWYERS, WHATEVER THEIR style of dress and whenever they send their faxes, have certainly been busy in Roberts County lately. Even a partial list of their activities is mind-boggling. In addition to the 1994 Quixx–CRMWA suit against the Panhandle GMD, which the city of Amarillo joined as an intervenor, there has been a suit by CRMWA against Mesa Water, and a suit by Mesa Water against CRMWA. There has been legal action by Mesa to force the groundwater district to grant permits, and legal action by CRMWA objecting to the language in the permits Mesa was eventually granted. A separate water-marketing group of Roberts County landowners has been put together by Amarillo attorney Ron Nickum; this group has traded offers and counteroffers with Mesa over potential lease agreements, so far with no conclusion.

In February 2003, Pickens filed an application on behalf of the Mesa Group to create a water district. The uproar that particular move set off could be heard all the way to the state capital in Austin; under Texas law, such districts are quasi-public bodies with the power to condemn rights-of-way for pipelines. Opponents immediately accused the former oil baron of seeking the condemnation powers to further expand his water domain, an idea that Pickens scoffs at. "There isn't anything we're going to condemn but a right-of-way 120 feet wide," he told me at the time. "And we can get that from the municipal purchaser of the water. We can get bonding authority from them also. The only reason I wanted it is that it gives us better leverage for negotiating with the end user of the water. Really, the people in Roberts County should *want* those things to be from the Roberts County end. But they act like we're somehow going to put some expense on them. It doesn't have anything to *do* with them. People make all these crazy statements about how we're gonna tax our neighbors, we're gonna condemn their water . . ." He snorts. "We can't condemn anybody's water." He has since withdrawn the application.

Pickens has expressed mystification as to why his project arouses so much ire. "There's no question I have surplus water," he complains. "If I had a surface that I could irrigate, nobody would question me irrigating corn. But the fact that I want to take the same water that I would irrigate corn with and sell it—that seems to bother some people." What he is doing, he suggests, will benefit the entire northeastern corner of the panhandle. "You've got four counties that are caught in this trap," he points out, referring to the counties that occupy the Canadian Breaks. "In those four counties, there are 2.4 million acres of land. Less than 100,000 acres of that are irrigated. If you could get a pipeline in there and sell the water, you could change the value of that land from $250 an acre to—who knows? Maybe as much as $750 an acre. That would be a huge, huge windfall. They say we're gonna sue our neighbors. We're not gonna sue anybody. All we really want to do is to change the value of the land in these four counties. Everybody can participate in the sale of water to a pipeline, over a period of years. But even if you don't sell, at least you can get your land appraised with a value for the water. Today you get no value for the water, because you have no place to sell it."

Brady is less puzzled by the attention. "It's because of who Pickens

is," he explains, driving along the edge of the Mesa Vista. "He's a controversial character. He came in to see us one time. I'd never met him
before. For all I knew, he was twelve feet tall, had horns, and a tail with a
barb on the end of it. Actually, he's a pretty pleasant fellow to visit with."

"I had a good conversation with him," I agree, "but I found him—"
I grope for an impression. "Found—"

"He has opinions," Brady supplies, mildly. "I don't think . . . I hope
I'm not being too naive. I don't think anybody really wants to drain all
of the water out of the panhandle. I don't think anybody involved in
this has that as some kind of evil goal. I do think there are a lot of egos
involved. You've got some pretty strong personalities. But I think that
everybody wants to do the right thing, to a certain extent. Of course,
some people might want to make a good profit while they're doing it."

AND THAT IS REALLY the crux of the problem. The issue isn't
really whether or not Roberts County water will be piped south: That
is already happening, through the CRMWA pipeline. It isn't a conflict
between irrigation and municipal water supply either: What little
irrigation exists in Roberts County lies well outside the zone of controversy. There may be legitimate environmental concerns, but these
are so far unproved, and if they show up they will probably halt the
project. Pickens is no environmentalist, but he does love his quail,
and he is unlikely to purposefully put them at risk. Controversies in
all areas save one, when examined, quickly evaporate. The one that
remains is profit.

Should water be an item of private profit? Is it ethically acceptable
to make money off something that is necessary for survival? If water
becomes commercialized, will the poor be literally priced to death? Is
an adequate supply of clean water a human right, or is it just another
saleable good?

These questions are not easy to answer, whatever your political
leanings. If you are of the liberal persuasion and want to make certain

that everyone has access to water at low or no cost, then you must deal with the tragedy of the commons—the tendency of individuals to abuse and overuse a free good. If you are a conservative and want to make certain that the right to make a profit from the property that you own is fully protected, then you must face the problem of the poor. All of the world's great religions stress that it is ethically necessary to take care of those who cannot care for themselves, and that certainly speaks against putting a price on water. But those who are most likely to stress the need to give free water to the poor are also those who emphasize the need to save our natural resources, and providing a good for free runs directly counter to encouraging its conservation. If it doesn't cost anything, why should you save it?

As both clean water and adequate government funding become scarcer, the question of private versus public water supply will continue to gain importance—not just over the Ogallala, but all over the planet. When water scarcity replaces water abundance, water's value as a commodity rises, making it more attractive to private providers; when governments face fiscal crises, capital-intensive utilities such as water distribution systems become harder to keep up. These trends encourage privatization. It is happening now. As of fall 2002, roughly 15 percent of American households were receiving water from privately operated utilities; that figure is expected to increase to 65 percent over the next twelve years. In Europe, the numbers are 40 percent today and 75 percent twelve years from now. Worldwide, supplying water to cities is a $300 billion industry that has attracted major corporate players, including Vivendi in France, Bechtel in the United States, and RWE in Germany. A good measure of the attractiveness of water to big-business strategists may be found in the tragicomedy that was Enron: One of its many interlocking subsidiaries, Azurix, was set up specifically to capitalize on the movement toward water privatization.

The privatization of water supplies is such an emotion-laden subject that it is difficult to find dispassionate information. Privatization advocates push it as a triumph of free-market entrepreneurship over government inefficiencies; antiprivatization activists see it as mixing the sacred (the water of life) with the unholy (profit). Stripped of their ideological fervor, histories of water-supply privatizations offer about the range you might expect: Some have been successful, some not. The most prominent bad example is Cochabamba, Bolivia, a beautiful old Spanish Colonial city of eight hundred thousand nestled at the base of the Cordillera Real. When its water system was contracted out to a subsidiary of Bechtel, in 1999, water rates exploded upward with no discernible improvement in the chronically poor service. That led to a series of riots since termed *la guerra del agua*, "the water war," in which at least one person, seventeen-year-old Victor Daza, was killed. Cochabamba officials were forced to withdraw in less than four years from what had been meant to be a forty-year commitment to its corporate partner. Bechtel has since sued the city for breach of contract. At the other extreme is Manila, the capital of the Philippines, where a well-thought-through process in 1997 succeeded in creating an award-winning private water utility that has managed both to improve reliability and to reduce costs for the five million people in its service area. The Manila Water Company has been so successful, in fact, that in 2003 a coalition of the region's poor petitioned the city to allow the company to *raise* its rates so it would have enough money to expand its infrastructure to serve more low-income neighborhoods.

In the United States, Atlanta's experience with privatization has more or less mirrored Cochabamba's, minus the riots and the tear gas; citizens in Georgia's largest city have complained of skyrocketing rates, poor service, and undrinkable water. On the other side of the country, in Burlingame, California, householders have been living contentedly with a private water system since 1972. With more than six hundred cities around the country under some form of water privatization, it's a safe bet that you will find both other Burlingames and

other Atlantas. Overall, it is probably an equally safe bet to say that
whether or not a water system is privately run has far less impact on
its quality of service than do a host of other factors, including the age
of its pipes, the competence of its management, and the quantity and
quality of the water supplied to the system to begin with.

That is not to say the debate over privatization is trivial. There is
much more at stake than whom you happen to pay your water bill to.
At issue is an entire philosophy of social welfare. Only the most rabid
misanthropes among us would deny that we all have an obligation to
maintain the integrity of our communities. Is that obligation collective or individual? Is it best encouraged through regulation, or
through fiscal policy? Are profit-taking and philanthropy mutually
exclusive, or does one make the other possible?

Water activists Maude Barlow and Tony Clarke leave no doubt about
where they stand. "We believe that fresh water belongs to the earth and
all species and that no one has the right to appropriate it for personal
profit," they write in the introduction to their 2002 book *Blue Gold*.
"Water is part of the world's heritage and must be preserved in the public domain for all time and protected by strong local, national, and
international law."

"People think water is free because it falls from the sky," counters
Andrew Seidel, president of USFilter, perhaps the most successful
American water-privatization company. "Well, it is. But treated, fil-
tered, and piped water isn't."

To Boone Pickens, the situation is straightforward. The water under
his ranch, like the ranch itself, belongs to him, and he should be the
one to control what happens to it. "To me it's so clear," he told me.
"I'm getting ready to be drained, or I *am* being drained, at this time. I
really have no choice other than to do what I'm doing. I'm not critical
of the Canadian River Municipal Water Authority; they bought water
and built a pipeline, and they're producing it. But if I don't offset *their*

production with *my* production, I'm going to let them take my water. And I don't want to do that."

Profit is not an issue for Pickens: He doesn't need the money. He has publicly offered to donate every penny he makes from Mesa Water to charity. That, of course, is the individual model of community responsibility. It fails to mollify his critics, who hold fast to the collective model and to whom, therefore, treating water as an item of commerce is heresy, no matter who might benefit from it. Pickens sees Mesa Water, Inc., as a specific response to a specific situation—a one-time deal. This, too, has failed to satisfy the critics, who cannot believe that he will stop at one deal, and for whom one is too many anyway.

"He's made a comment that getting this going is harder than any project he's ever done," remarks Ray Brady. "I'm not sure what that means."

I suggest that it means the groundwater district has been effective. Brady shakes his head. "I don't think that's what we're doing," he says. "We're not intentionally trying to make things difficult. I hope what we're trying to do is to make sure that everybody has an equal opportunity, and that we don't do something to totally deplete everything."

We have left Mesa Vista far behind now, and are somewhere in Armstrong County, up on the Llano Estacado, traveling south toward Palo Duro Canyon. The land is flat as a table, and the center pivots are shoulder to shoulder. Pickens is right, I am thinking: If these guys can irrigate with their water, then he should be able to sell his. The effect on the resource is identical. Whether or not the effect is acceptable is a separate question, and that question should be addressed in the context of all uses, not any one alone. Aloud, I mention to Brady that people often have trouble distinguishing the actual effects of their actions from the desired effects. Brady nods.

"Another thing that seems to get lost is the time," he says. "What kind of time period are you looking at? Are you looking at a ten-year period? A fifty-year period? A hundred-year period? A ten-thousand-year period? What's your frame of reference here? That can make a tremendous difference in what you see and how you talk about it."

An ancient rural fire truck comes toward us in the opposite lane, lights on and siren blaring but going only about forty-five miles per hour. "Too heavy to go any faster," Brady observes. "He's got a load of water—8.34 pounds per gallon. And the other number's 7.48 gallons per cubic foot. Gotta know those—you can always tell the real water people."

"In Europe, they have it easy," remarks my wife, "with the metric system."

"Yeah," agrees Brady. "ten-ten-ten-ten. All right, but that gets boring after a while. How many *varas* is it? Anyway, here's a little preview of things to come." Palo Duro Canyon has suddenly opened before us, a great Kodachrome slash in the Llano Estacado, several miles across and nearly a thousand feet deep. After all the flatness, looking over the edge is a stomach-dropping sensation. There is no gradual transition, as when you enter the Canadian breaks: One moment the ground is level, the next it is vertical. The walls are brightly banded, white caliche and pinkish-tan Ogallala and, at the bottom, the brilliant, cardinal-red, Permian-era formations known as the Spanish Skirts.

The road finds a way down the impossible wall. "This is a little bit bigger than your average west Texas gully," comments Brady.

We are all silent for a while. When Brady speaks again, it is to pick up the conversation where it had been before the fire truck interrupted it. "I think it's hard to find anything absolutely black or absolutely white in real life," he muses. "Where's the balance that gets the best return for the brainpower and time and money that's invested? Where is that? To me that means you have to do continual self-evaluation and feedback. And I don't want to sound like a Communist cell member, or anything like that. But I think that self-assessments are important. You have to evaluate performance and ask, 'Are we doing the right thing?' I'm not sure what's going to happen up here. My feeling is that forces other than the amount of water are going to have a greater impact on water than anything else."

"Do you think you'll be able to hold the 50/50?" I ask.

"I think in some places we can," he says cautiously. "I think we will be challenged in others. I'm not sure that, hydrologically speaking, we can always justify consumptive restrictions. I think for the majority of the district, we're okay. There are some isolated areas where it's probably not going to be possible. Where those areas are may not depend on what the rule is, or whether or not we're enforcing it."

"And then the next question is, what happens after fifty years?"

He nods. "That could be interesting. The struggle to get the first fifty years was so horrendous, I'm not sure anyone wants to get into that next part. But it's something we have to work on. Are people willing to even do what we're currently doing? Is that a viable goal? It may be that the voting population says, 'No, that's not going to work. We're gonna throw all these directors out, and we're gonna get a new set of directors that'll let us pump however much we want, whenever we want.' That's always a possibility. And I really don't think the municipal population cares what the groundwater level is, as long as you turn on the tap and the water flows. So water conservation becomes not so much a regulatory function as an educational function. The way you have to approach it is, 'If you don't quit wasting water, you're not going to have any when you turn it on.' And if you can show that municipal withdrawal is going to affect that directly, that's fine. But we have a difficult time doing that here, because over 90 percent of our groundwater goes to agriculture.

"Of course, the traditional rallying cry is, 'Do you want to eat?' Well, most Americans don't eat cotton, and they don't eat milo. They eat corn, but most of this corn doesn't go to the table, it goes to the feedlot. So all these things will have an effect. I think the biggest one is education, more than anything else. But that brings up the question, What do you teach?"

What values does the community want its children to learn? Community responsibility, certainly, but will it be collective responsibility or individual responsibility? Who decides? As water becomes scarcer, on the High Plains and elsewhere, these questions assume more importance. Pragmatically, there are several correct ways to answer them. Ideologically, there rarely seems to be more than one, and that one often seems diametrically opposed to the one your neighbor demands that you accept.

Brady is talking about the groundwater district's education program. "We have a guy who teaches water conservation," he explains. "He goes around and does fifth-grade classes. He's talked to over twenty-five hundred fifth graders this year. Is that the right way to do it? Well, you can argue that, too, like everything. But I think conservation's going to be more and more important to us."

Conservation is important to Boone Pickens, too, but he would probably define it differently than Brady does. The traditional definition of conservation—the definition pioneered by early twentieth-century conservationists such as Gifford Pinchot and Teddy Roosevelt —has to do with making sure that there is always enough to go around, for us and for future generations. But conservation can also be viewed as making certain that nothing of value is wasted, and that, I think, is what Pickens is trying to do. The water under his land has value. It would be a waste to let that value go without trying to reclaim some of it. That may not match Brady's definition of conservation, or yours, or mine. Selling water in order to save it may seem an unlikely course. But this is Texas, where the Rule of Capture still holds. Preserving the value of water may be an inferior choice to preserving the water itself, but it may also be the only choice available.

There is no question that Pickens thinks so. "I'm not mad at the Canadian River Municipal Water Authority," he insists. "It isn't a case where I think those guys shouldn't be doing that to me. I don't look at it that way at all. The way I view it is that I've got to protect myself. I place no blame on them. It's up to me."

We have reached the end of the conversation. Phone numbers and e-mail addresses have been exchanged. Pickens hesitates. He clearly wants to make certain he has been understood. "Just remember, I said it," he tells me, finally. "*I'm gonna sell the water.*" He laughs. "I'll be damned if I'm gonna let somebody take it away from me."

XXIII

DROWNING TEXAS TECH

WHEN MORE AND MORE people compete for pieces of a smaller and smaller pie, contention always escalates. The law of supply and demand guarantees this. If the High Plains states don't sue each other over the Republican River, then it will be the North Platte, or the Canadian, or the Arkansas; if a pipeline is prevented in Texas, then a bottled-water plant will bob up in South Dakota; if Boone Pickens backs out in Dallas, then an entrepreneur in Denver or Kansas City or Omaha will step in. It is only a matter of time before New Mexico and Oklahoma take Texas to court over the Rule of Capture, or before the thirsty cities along the Front Range of the Rockies exercise the combined muscle of political clout and money to gain access to the water beneath the Sand Hills. The Ogallala is a diminishing resource, and potential uses for it are on the rise. No one should be surprised that tempers are on the rise as well, or that politicians, judges, and activists are locked in a seemingly endless round of more laws and more lawsuits as they attempt, like Solomon, to divide the baby without cutting it in half.

But here another possibility suggests itself. Water is renewable; aquifers are capable of recharge. The Ogallala's recharge is very slow, but it is not nonexistent. Could it be sped up? Could the solution to a shrinking pie be not smaller pieces, but a larger pie?

In theory, the answer to that question is yes. In practice, the only successes have come where no one suspected they would—and no one particularly wanted them to, either.

RAY BRADY BEGAN running in the army in 1966. He has been pounding pavement almost daily ever since. "I was never a track runner, just a long, slow plodder," he says today. His mileage has extended over the years: two miles, five miles, ten miles. In Houston in the early 1980s he began to run marathons, training at noon on the paved paths along the bayous with the humid breeze providing "plenty of sauna effect." He has encountered rattlesnakes, pheasants, lizards, deer, and—in south Texas—illegal immigrants. On active duty in Korea, running on a path parallel to the DMZ, he had to plan his route carefully and hold strictly to it to avoid straying into minefields. Perhaps his oddest encounters, though, came during study-break runs at Texas Tech University while he was attending graduate school there in the early 1990s.

"I would go running by Jones Stadium," he recalls, "and they would be pumping water out from under the stands. It would be running down the street." He shakes his head incredulously. "Here we are, in west Texas, and we're dewatering the football field."

Jones SBC Stadium looks like a Spanish hacienda on steroids. The home of Texas Tech's Red Raiders stands at the north end of the university's campus in north-central Lubbock, hard by the largest air-inflated domed building in the world, which is used by the team as a practice facility in the unlikely event that it should actually rain on the Texas Panhandle. The stadium, which used to be just plain Jones Stadium before SBC Communications coughed up $20 million toward its remodeling fund, can hold up to fifty thousand screaming Tech fans, almost none of whom will be aware of the leaky basement. Jones Stadium leaks from the outside in. Here in dry Lubbock, the commercial center of the Llano Estacado, amid dropping water tables and failing wells that have been forcing farmers out of buisiness since the early 1970s, Texas Tech is drowning. There is much more water beneath the university than it wants.

The Tech campus has a scholarly, ivied look. Buildings of blonde brick dating from several eras of academic architecture rise at comfortable distances from one another across green, tree-shaded lawns. Automobiles thread narrow drives that lead to large parking lots located discretely away from the central quad. The lawns and the parking lots are part of the problem. The lawns are watered regularly, and all water in excess of infiltration and evapotranspiration runs off into the parking lots. The paved parking lots form impervious surfaces from which water drains; the runoff now goes into Lubbock's storm sewers. The sewers convey it to the convenient playas scattered throughout the city. The once-dry playas are now wet playas: after many decades of use as storm sumps, most of them hold water year-round. As the city has grown and more stormwater capacity has been needed, bulldozers have deepened many of the playas, digging through their clay bottoms and increasing their infiltration capacity. If Lubbock had set out to design an aquifer-recharge mechanism instead of a system for getting rid of storm water, it couldn't have done a much better job.

Had it stopped at that, though, the stadium—and the Art-Architecture Building, and the Business Administration Building, and the fancy raquetball court donated to the university a few years ago by H. Ross Perot—would not require regular pumping. Two additional factors have come into play. The first is Lubbock's growing thirst; the second is a spell of dry climate that the Llano Estacado endured during late Ogallala time, more than five million years ago. The contemporary thirst has brought extra water into the city. The ancient drought has left the extra water with no place to go.

The water comes in through Lubbock's municipal water system, which draws it from Lake Meredith, on the Canadian, and from a well field near Muleshoe, on the Texas–New Mexico border. The drainage barrier is an all-but-impervious caliche layer formed by the evaporation of mineral-rich waters from the Llano Estacado's surface during that spell of dry weather five million years ago. It is now buried well down in the earth, but its presence prevents any water that sinks through the bottoms of the playas from making its way down to the Ogallala; it remains on top of the caliche, forming what is known as a perched aquifer. It is this perched aquifer that is creating the university's problem. Water drained from beneath New Mexico and water sucked out from under Boone Pickens's ranch is pooling on the caliche and pouring into basements at Texas Tech.

Well, isn't recharge a good thing? The principal problem facing the Ogallala is drawdown, which results from withdrawals in excess of recharge. Shouldn't we be happy that recharge is exceeding withdrawals in Lubbock? The perched aquifer beneath Texas Tech has been estimated to contain an eleven-year water supply for the city, at current rates of consumption. Flooded basements would seem a small price to pay.

But there is a larger question to be asked. Is what is happening beneath Texas Tech duplicable elsewhere? Can the circumstances that produced the groundwater mound Lubbock does not want be recreated someplace else to produce a groundwater mound someone does want?

BACK TO GROUNDWATER 101. An aquifer consists of an underground body of sand, gravel, or other fractured material that contains gaps large enough to hold and transmit water. Aquifers are characterized by *porosity* and *permeability*, meaning that they contain many small spaces (pores) which connect to one another (are permeable). The greater the porosity, the more water the aquifer can hold; the greater the permeability, the more rapidly the water will move.

Water relates to materials in four different ways. *Free water* (sometimes called "gravitational water") flows freely over, under, around, and through things in response to gravity. *Film water* clings to the surfaces of things: It is what makes you stay wet after you have turned off the shower or climbed out of the tub. *Capillary water* is held in very narrow spaces by the phenomenon of surface tension: It refuses to drain out, and may actually climb upward, in apparent defiance of gravity. And *bound water*, as its name implies, is water that is bonded directly into the physical structure of another substance. It contributes to the characteristics of the material to which it is bound, but it also retains its identity as water and will slowly evaporate over time (think of the difference between fresh and stale bread).

Aquifers have two zones: a *saturated zone* (or "zone of saturation"), where all of the pores are full of water, and an *unsaturated zone* (or "zone of aereation"), where some or all of the pores are filled with air. The boundary between these two zones is the water table. To recharge an aquifer, you must add to the volume of the saturated zone, which means that water must move down through the unsaturated zone to the water table; of the four types of water just cited, only free water is able to do that. Bound water, film water, and capillary water don't move. They are available to plants, and they may be removed through evaporation, but they will never reach the water table and they cannot help refill wells or keep springs alive.

So the problem of artificial recharge may be redefined as the problem of passing enough free water through the unsaturated zone to raise the water table. There are two ways to do this. One is to inject water directly into the saturated zone; this requires pushing it down wells under pressure, consuming large amounts of energy and risking well-casing ruptures. No cost-effective method for accomplishing recharge in this manner has been found. The other way is to introduce enough water into the unsaturated zone so that after some of it becomes bound water, film water, and capillary water enough will remain as free water to make an impact on the water table. This is nature's approach. In order to be successful, any artificial-recharge scheme will have to find a way to improve on nature's methods.

The principal recharge mechanism for the southern part of the Ogallala Aquifer is the playa lakes. These low spots in the otherwise flat High Plains topography collect runoff from storms and hold it for periods ranging from several hours to several weeks. Playas tend to have clay floors, and for that reason they were discounted as recharge mechanisms for many years. In the late twentieth century, though, geologists began looking at them more closely, and when they did, they discovered two things. One was that the clay is usually split by small cracks that enlarge as water drains through them, a process called "piping." The other was that the clay usually doesn't extend all the way to the edge of the playa. There is a ring around the inside of the rim, a doughnut-shaped zone of sandy soil, through which infiltration can happen quite rapidly. And because playas are inundated off and on throughout the year, they do not develop much vegetation, so there is not much competition from plants for the water. Studies suggest that roughly half the water that enters a playa will infiltrate, accounting for 95 percent of all aquifer recharge for the southern High Plains.

Where playas do not exist, the majority of recharge to the Ogallala takes place through sand dunes. Sand creates almost no runoff: nearly all of the rain that falls on a dune sinks straight down and plunges toward the water table. Dune vegetation is even scantier than playa vegetation, so there is little evapotranspiration to interrupt that downward journey. A small amount of water is retained in the upper part of the sand, to replace film water and capillary water that has been lost to evaporation; the rest rapidly enters the aquifer. This is the process that has created the huge reservoir beneath the Nebraska Sand Hills, and the remarkably steady flow of the streams that originate in that area.

A third important route of recharge is infiltration through riverbeds. This takes place primarily during storms, which are the only times that most High Plains rivers have water in them. The process is similar to playa recharge, but because much of the water escapes downstream, less is available for infiltration. Studies of ephemeral streams in western Kansas suggest an infiltration rate of 1.3 percent of total stream flow per mile of streambed. If that figure is accurate, it would take forty miles of ephemeral stream to equal the recharge from a single playa a quarter-mile across.

UNDERSTANDING HOW NATURE recharges aquifers is the first step toward artificially augmenting it. The second step is harder: You must discover something that nature missed. Attempts have been made, but the results have not been particularly encouraging.

In Texas, artificial recharge attempts have concentrated on the playas. Two principal approaches have been employed. One is to introduce additional water; the other is to alter the playas' clay bottoms, either by scoring them to increase piping, or removing the clay altogether to allow water to infiltrate through the entire bed. Usually the two approaches are combined: extra water plus improved infiltration. This is what Lubbock has done accidentally. It has not worked quite as well where it has been applied on purpose.

"I haven't seen any evidence that they were wildly successful," remarks Lubbock's Dana Porter, who once worked under one of the artificial-recharge researchers. "It hasn't been as easy as they thought it would be. It's very energy intensive."

Increasing recharge through the playas "didn't work out too well," agrees Ray Brady. A significant part of the problem, he believes, is the much greater depth to groundwater on the Llano Estacado when compared to the perched aquifer beneath Lubbock. The thicker the zone of aereation, the more the water that might have gone into recharge will get sidetracked instead into bound water, film water, and capillary water. "Getting water to the water table takes time and distance," Brady points out. "It's not rocket science; it's just time and distance. If water has to infiltrate three hundred feet, you're just not going to get as much down there as if you only had to go two feet."

In northeastern Colorado, where there are no playas, attempts have been made to increase recharge by damming gullies. The theory is that if storm runoff is held back, it will have time to infiltrate. Again, there has been something of a gap between concept and reality.

"There have been a number of projects over the years to try to improve recharge through the intermittent streams, but nothing was ever really successful," explains Wayne Shawcroft, the Akron extension-agent-turned-banker. "They'd put these check dams in the waterways, and they'd go maybe five or six years with just a trickle. Then all of a sudden there'd come a flood, and it would wipe out all of the check dams." He chuckles. "They finally figured if it was that big of a flood, it would have recharged anyway, no matter what they did."

It was hard to sell the research projects to grant providers, because it was hard to determine whether or not they were effective. "And if you use runoff water for recharge," Shawcroft points out, "that's not as good, quality-wise, as what's already down there. Runoff tends to accumulate a little sediment, and maybe some chemicals and things—why put those down into the good-quality water? And then you just have to pump it back out again. So you wonder whether you're really doing an effective job."

You will notice that Shawcroft emphasizes quality problems. That is also the difficulty with using Lubbock's groundwater mound as a water source for the city. It was created largely by percolation from pavement runoff and oversprinkled lawns, and an assay of its contents is not encouraging. "It's very low quality," says Porter. "All those wonderful lawn chemicals. Don't look at my lawn, it's not the prettiest. I'm the water and conservation guy; I'm not going to have the greenest lawn in the neighborhood." Natural recharge suffers from some of the same quality problems, but it does not suffer them to the same degree. Artificial recharge schemes all depend on concentrating runoff, which also concentrates any contaminants. Natural runoff is more dispersed, so chemicals remain lower percentages of the total. And even where natural recharge has introduced serious quantities of harmful chemicals into the aquifer, one would still have to question the wisdom of deliberately adding more via artificial recharge. When you have a polluted source that you are attempting to clean up and use, pouring more pollutants into it does not seem a particularly sensible way to go.

Along with quality issues and mechanical problems, attempts at artificial recharge have run into legal roadblocks. The area over the Ogallala in Colorado, for example, is tributary to the South Platte, the Republican, and the Arkansas rivers. Any water that is held back by check dams so that it infiltrates instead of reaching those rivers is water that must be accounted for, somehow, in the interstate compacts that govern them. And even if there weren't claims on that water from Kansas and Nebraska, there would still be claims to deal with in Colorado itself. Collecting water for recharge impedes the flow of water that has already been appropriated. On the water-poor High Plains, water rights have been filed nearly everywhere water might possibly exist, including most of the intermittent streams. Under prior appropriation rules, those filings take precedence over using the water to recharge somebody else's well.

FINALLY, IN COLORADO and elsewhere, those who would increase recharge run into simple arithmetic. To put water into the ground, you must have water available. Where will that water come from? If rainfall and stream flow are inadequate for growing crops directly, how can it help to push the water into the ground and then pump it back out again later? Ultimately, the only answer to inadequate water supplies is more water. How do you get more water to this parched and arid land?

In the late 1970s, when the issue of water table decline first arose, the answer to that question tended to be "pipelines." Were the High Plains running out of water? Then go get some. Replace lost Ogallala water with water from elsewhere, so the rich soils of the plains could continue to produce crops. Where the replacement water would come from didn't seem to matter, as long as it was someplace with a surplus.

Promoters in Texas and Oklahoma hyped a pipeline from the lower Mississippi River. The Mississippi pours 390 billion gallons of water into the Gulf of Mexico every day; why not trap some of that and send it west? Cooler heads pointed out that this would require pumping the water more than three thousand feet uphill. A gallon of Mississippi water delivered to Lubbock would require the expenditure of at least twenty-five thousand foot-pounds of energy. At that rate, each acre-foot would cost the equivalent of eighty-three gallons of gasoline. An acre-foot pumped from a really deep well—a thousand feet—costs just twenty-four gallons. The plans were shelved.

An even grander scheme, the North American Water and Power Alliance (NAWAPA), proposed nothing short of replumbing the entire continent. Northward-flowing rivers in Canada would be dammed near their mouths on the Arctic Ocean, and the water collected behind these dams would be pumped south. The Great Lakes would become a staging area; a second staging area would be developed in western Canada by damming both ends of the five-hundred-mile long Rocky Mountain Trench. Tunnels would be drilled under obstacles, including—in one version—most of the state of Idaho. The payoff would be more water for the High Plains, for the parched Sun Belt cities in California and Arizona, and—magnanimously—for the farmers of northern Mexico, who were complaining that American demands on the Rio Grande and the Colorado River weren't allowing adequate amounts of water to flow across the border. NAWAPA would have cost billions to construct and billions more to operate, and would have displaced thousands of people and jiggled the continent's climate in unpredictable ways, and although there are still people today who believe it should be built, it too has effectively been shelved. In fact, pipelines in general have been shelved. With the exception of peripheral projects, such as South Dakota's Mni Wiconi pipeline, hauling water to the High Plains via surface transport is no longer considered a viable alternative.

There are, however, other ways to try to get it there.

"HERE'S WHERE WE KEEP OUR AIR FORCE," says Ray Brady, waving a hand toward Perry Lefors Field. We are headed south on Ranch Road 282 just outside Pampa, Texas, passing the little Gray County airfield along its eastern flank. "We just moved one plane over to Amarillo," Brady continues, "because one of our pilots lives there. He can get to his plane quickly if things develop favorably for cloud seeding." Cloud seeding is another name for weather modification, which is a somewhat less snicker-prone term than rainmaking. The Panhandle Groundwater Management District is staking part of its future on silver iodide crystals and hope.

"You've got two pilots?" I ask.

Brady nods. "We've got two pilots, two planes, two of everything. We've got a twin-engine plane and a single-engine; both of them have been used for a number of years in other programs." The advantages of buying a plane already set up to do cloud seeding go beyond merely saving the costs of conversion. "If you modify a plane, then you have to get the FAA to approve it, and there are a lot of inspections and regulations and expenses before you can do that. If you change anything, the FAA has to certify airworthiness." Brady smiles, a bit ruefully. "You learn a little bit about a lot of things in the groundwater business."

The Panhandle GMD is not the only Ogallala entity that has been involved in cloud seeding. The Lea County Water Users Association has tried their hand at it in New Mexico, as has the Great Bend GMD in central Kansas. In Lubbock, the High Plains Underground Water Conservation District began a weather modification program but had to shut it down midseason in the face of "very vocal opposition" from some of its constituents. What kind of opposition? Brady smiles. "Some of it is opposition on a religious basis: If God wants it to rain, it'll rain; man should not experiment. There is some of that. In this particular case, though, I honestly don't know what the detractors wanted. My personal opinion is that some folks have gotten pretty powerful moneywise, and are thinking they ought to have more say than others." He clams up—good-naturedly, but determinedly—when I press for more.

The oldest Ogallala cloud-seeding program—one of the oldest anywhere—is at the extreme south end of the aquifer, in Big Spring, Texas. Chris Wingert, the assistant manager of the Colorado River Municipal Water District, talked to me about it on a bright April morning with not a cloud anywhere in the sky to squeeze a drop of rain out of.

"We've got the longest-running program in the state of Texas," Wingert told me proudly. "We actually started back about 1971. The district had just completed the E. V. Spence Reservoir down in Coke County, but the lake hadn't caught any water yet, so we were really scrambling to try to boost our water supply for the short term. Unfortunately, you cannot fill up your lakes off a weather modification program. If we could, we'd be in a lot better shape."

He explained the rainmaking process. Like most weather modification attempts, Wingert's crew uses cloud-base seeding. The pilot flies a modified Piper Aztec along the base of a cloud, looking for updrafts. When he encounters one, he ignites silver iodide flares mounted to the trailing edges of the wings; the updraft carries the silver iodide crystals to the top of the cloud, where they become nucleii for raindrops. "There's free moisture up in that area," Wingert noted, "but unless it's got something to condense around, it just remains free moisture. So what we're doing is providing more things for raindrops to condense around."

I asked if they had any data on the efficacy of the program. Wingert said that they had. "We've done extensive studies on the improvement in rainfall," he stated. "We've looked at dryland cotton yield within the target area, upwind of the target area, and downwind of the target area. What we see is about a 20 percent increase in production in the target, compared to upwind and downwind. Another thing we look at is rain gauges. We have a network of about eighty-five fence-post gauges scattered out along highways, and our meteorologist checks them after every storm. Some of these are within our target area, and some of them are outside the target area, so we get a comparison. And with those rain gauges, as with the cotton, we see a similar 20 percent difference."

Twenty percent more rainfall does not necessarily translate into a 20 percent improvement in water availability. "What makes it kind of hard," admits Wingert, "is that while you might get 20 percent more water on the ground, what happens to that water? Some of it ponds up and evaporates, so you don't get anything there. Some of it goes into the groundwater. A lot of it is used by plants, which is why we're seeing the increase in cotton. And then some portion actually does meander into a stream and make it down into the lake. Are we seeing a 20 percent increase in production in our lakes? No. We don't have any we can point to. Are we seeing an increase in the water in the lakes? Yes, we feel like we are. It's relatively cheap insurance. Even if you only get five to ten thousand acre-feet per year, that's some of the cheapest water you'll buy. It gives you another tool in your toolbox. Rainmaking is just another way to wring out every drop that we can in this part of the country."

There are many who hold doubts about weather modification.

Dana Porter is one of them. Not because she is convinced that it doesn't work, but because she fears that no one can really tell when it does, meaning that rainmaking programs will always remain on shaky financial ground. "Some efforts have been made to study it," she states. "They'll seed some clouds and not others, for example. But it seems like a very difficult experimental design to impose. You have to find comparable clouds, and they have to be over comparable areas, with comparable drought conditions. It's hard for me to get my small mind around what they have to do to get fair treatment of the data."

Ray Brady understands this attitude, but like Chris Wingert he remains cautiously committed to the program. "Our theory is that groundwater is going to be our resource," he explains in Amarillo. "But groundwater is pretty much finite in our district. So you can either recharge, or you can try to make it rain a little bit more and cut the demand. Is this viable? Are we doing something reasonable? We don't really know. But we're always complaining about rain. What does it hurt to try? We'll prove it one way or the other, and then we'll go from there."

XXIV

THE LONG FAREWELL

I F NO WAY IS FOUND to stop the decline of the Ogallala Aquifer, what will happen? The answer is brutally obvious: The water will run out. At that point there will be no choice but to get along without it. High Plains residents will farm dryland—without irrigation—or they will quit farming altogether.

Will it work?

Jeff Johnson has been looking closely at that question. Johnson is a doctoral candidate in agricultural economics at Texas Tech University, but he is no dewy-eyed ivory-tower youngster; he worked as an extension agent in various parts of Texas prior to his stint at Tech, and his original association with the university came not as a student, but as Director of Farm Operations.

"I just kind of stumble-bumbled into the water issue," he told me in his second-floor office in Tech's old Agricultural Sciences building. "When I took over the farm it was, of course, 100 percent irrigated. So I had to learn about that, and when I started seeing all of the interesting issues that go on around this aquifer, I just got excited about it. I thought as long as I was here, I might as well go ahead and get my Ph.D. In all my spare time."

For his thesis, Johnson has been looking at the economic changes that a conversion to dryland will bring to the southern High Plains. What he has been finding is a tendency toward consolidation. "As the Ogallala declines and we go to dryland, the farms will have to get larger to grow the same amount of crop," he pointed out. "You've just got to have more acres. If you have larger farms, you have fewer people. What happens, then, to the economy? When are we going to start to lose small towns? When do we end up being Cattle City, like some of those counties out in west Texas?"

I asked him when to expect the collapse. He shook his head. "I don't expect it to be a collapse," he said. "It's going to be a long, slow decline to a lower level, and it'll be a generation before anything noticeable happens. It's not going to be all of a sudden people are immediately vacating the area." Water table decline is an incremental process, and irrigation declines incrementally with it. The crucial factor is not static level, but well yield. As the water gets thinner, yield drops; as yield drops, fewer acres can be irrigated. Little by little, the farm reverts.

"Say I've got a place out here," Johnson explained, by way of illustration. "I'm irrigating 320 acres. As the water table declines, and my water starts getting less, I'm not going to stop irrigating, I'm just going to cut my irrigation to 240 acres. Or 200 acres, or 100 acres. And when it gets down to where I don't have enough water to irrigate *any* acres, then I'm completely dryland. But that's going to be a matter of years."

Local variations in saturated thickness and in permeability—the principal factors that affect well yield—will cause the yield to decline at different rates in different wells. Crop choices, management practices, and localized drought will all contribute to the pattern. The High Plains will not lose irrigation all at once. "It'll be a soft landing," Johnson emphasized. "But I do see it going dryland eventually, and when it does, farm size will increase. In the past, a 640-acre farm, or a thousand-acre farm, was a good-sized spread. I see it going to two or three thousand acres. Dryland, mostly. There'll still be some irrigation, but they'll put it on high-value crops on small acreages." The Ogallala will not go away entirely; it will just become very thin, very low-yield, and very, very dear.

Increased farm size does not necessarily imply a shift to corporate farming. "I think what you'll find is what you're beginning to see right now," Johnson stated. "There are individual producers who have five thousand, six thousand, even eight thousand acres, and they hire two or three people to help them. I think we're going to see more of that. If that's agribusiness, so be it. But it's not that much different from the structure and organization of the businesses we already have, it's just a heck of a lot more acres."

The business structure may stay the same, but the larger farms are likely to operate differently. There will be several equipment barns scattered around the property in place of the single barn that is common today. Expensive machinery such as combines and mechanical planters may have to be duplicated. "They'll have, probably, more tractors," Johnson observed. "They're not going to be able to road their tractors back and forth on these huge acreages." Capitalization will increase. Farming—already a financially risky occupation—could become riskier.

I asked if the changes were likely to affect Lubbock. Johnson nodded. "Lubbock is a central point for most of the twenty-three counties I'm looking at, so I'd say it will affect Lubbock some. But it may be a third-order effect. Lubbock's diverse—it's got the medical community and the university. Some of the businesses that depend on cotton will be affected, and they're a significant chunk of the economy. But Lubbock is not going to fold up."

"Is dryland cotton an effective crop around here?"

"Well, you're not going to get the yields. But you're not going to have the inputs, either. It's a low-input, low-cost crop, and everybody's done enough dryland cotton that they know how to manage for it. Yeah, dryland cotton around here is a good crop."

————

TEN MILES AWAY, at the Lubbock Extension Service Center north of the airport, I got a slightly different take on High Plains dryland farming from Dana Porter. I began by asking if she thought the Ogallala would survive in some form. She shook her head. "At some point, my guess is that it'll no longer be economically advantageous to use it," she stated. "It'll be harder to pump, and more expensive. It's a sponge, you know, and I think it will deplete. Economics is going to drive a lot of that."

"What's going to happen then?" I asked. "Will there be a successful conversion to dryland?"

"There will be a conversion to dryland."

"I notice you left the word 'successful' out of there."

She smiled. "Well, it's much higher risk. Around here, we might get rain in the spring, or we might not. If we don't get rain in the spring, we won't have a crop, period. If we get rain in the spring but not much in the summer, we'll have some crop, but it will be of much lower value. You can't count on having a crop every year. It's hard to explain that to people who live in areas that have a little more rainfall. They want to know why we don't just go to dryland. Well we could, but we'll have a crop maybe one out of every three years, or one out of every five years."

"What I've heard people ask is, 'Why don't they farm Ohio, where it rains by itself?'"

Porter snorted. "Yeah. Just move there. There are reasons why we don't. Cotton and peanuts don't grow well in Ohio. They have a lot of disease issues with peanuts in the southeast. We're able to grow them here less expensively with fewer chemicals, despite the fact that we have to pump a lot more water than they do. So if I have a peanut-butter sandwich, my rights to complain are limited."

Like Johnson, Porter expected a negative impact on small towns. "When you switch to dryland, you're going to see a decline in a lot of pasture covering most of the Great Plains. Declining water tables d declining populations, they argue, demonstrate that farming is t sustainable here. "Tear down the fences, replant the shortgrass, nd restock the animals, including the buffalo," they wrote in the 1987 aper in which they first proposed the idea. "The federal government's commanding task on the Plains for the next century will be to recreate the nineteenth century."

In many areas, the distinctions between the present national parks, grasslands, grazing lands, wildlife refuges, forests, Indian lands, and their state counterparts will largely dissolve. The small cities of the Plains will amount to urban islands in a shortgrass sea. The Buffalo Commons will become the world's largest historic preservation project, the ultimate national park. Most of the Great Plains will become what all of the United States once was—a vast land mass, largely empty and unexploited.

Predictably, the concept was vilified. The Buffalo Commons, wrote then-governor Mike Hayden of Kansas, made "about as much sense as suggesting we seal off our declining urban areas and preserve them as a museum of twentieth-century architecture." (Hayden has since come around to a position of cautious support.) Angry ranchers showed up at meetings to denounce the idea and its promoters. The Poppers insisted they were misunderstood. All they were saying, Deborah Popper told the Lawrence, Kansas, *Journal-World* in early 2004, "is that when the Plan A Economy—that is, agriculture—fails, there needs to be a Plan B Economy" based on ecologically sound principals. They have called the idea of the commons a metaphor that got out of control. "Early on, we were accused of being for some sort of forced, government-led land grab," Frank Popper complains. "I have no idea where that came from."

How much sense the Buffalo Commons idea makes depends very

the communities," she emphasized. "That's

think it will accelerate. There are some at Te

returning to the grassland and grazing system.

and windmills. We could be very romantic about

"Should we replace the cows with buffalo, bec

suited to this landscape?"

"No."

"Some people think—"

"'Cause the song says this is where the buffalo roam.

"In some areas it's easier than it is in others. The manure

lem. If I just put it on dry ground and leave it out there, i

dehydrate, and then when it rains, I've got reconstituted . .

the sentence hang. "Or I could use it as a soil amendment fer

I'm growing a crop. But that's got to be managed according t

nomic requirements. It can be done, if it's done carefully. I think

are people who do a good job of it. There are some who don't.

question is how to manage the overall system. How to spread the b

falo out far and wide."

"If they actually roam," I suggested.

"If they actually roam," she agreed. "If there aren't things that

inhibit them from doing that. The same thing holds with cows, by the

way. There are ways to distribute their water tanks and control the

traffic. Encourage them to move to the other side of the field. Rota-

tional grazing. It's a matter of shifting the animals in a controlled pat-

tern. They condition to it very well, because they're always moving to

better grass."

Though she did not mention them by name, Porter was undoubt-

edly thinking of Frank and Deborah Popper as she responded to my

buffalo question. The Poppers hail from New Jersey: he teaches land-

use planning at Rutgers University; she is a geography professor at the

Staten Island campus of the City University of New York. They have

proposed what they call a "Buffalo Commons," a vast, fenceless buf-

much on who you are and where you are. There is no question that
as water for irrigation disappears, grazing will take on increased
importance. Which animal? Buffalo are a good candidate, but so are
longhorn cattle, a breed that thrives under dryland conditions. Wild
herds of both species roam the Fort Niobrara National Wildlife
Refuge atop the Ogallala Aquifer in northern Nebraska apparently
with equal success. Regular cattle, as Porter indicated, may also do
well with proper management. And what about pronghorns? These
"antelope" (actually, distant relatives of goats and sheep) are as
much at home on the plains as are buffalo. They regularly graze
among cattle. There is no reason that they could not be raised as an
adjunct to cattle ranching.

But any of these animals may have trouble surviving on the post-
Ogallala High Plains. In presettlement times, buffalo and pronghorns
depended on Ogallala-fed springs. Many of them have stopped flowing, victims of the same water table decline that farmers are struggling
with. There will be enough water to support some animals, but certainly not in the numbers recorded by early observers. No matter what
we do, the great herds are a thing of the past.

The whole argument over which grazing animal is best for the High
Plains tends to exasperate Trudy Ecoffey, the South Dakota range ecologist. "I wish people would see the buffalo not as competition, but just
as a different animal," she told me. "People in ranching usually see it as
a threat. I think there's enough room, and there's enough people eating meat out there, for both. If you make it sustainable, maybe you
can get some revenue off it." She is currently involved in ecological
studies on Lakota lands, trying to determine the varying impacts that
cows and buffalo have on the plains environment. The differences she
is beginning to uncover may help determine the best places for each.
This could bring about what the Poppers say they really meant to suggest: an evolved patchwork of buffalo commons, cattle pasture, and
farms, each fitting neatly into its own agricultural niche.

———————

AND A PATCHWORK is what we are likely to get, anyway. There is no one-size-fits-all approach. Technology can squeeze more production out of declining water, but it cannot stop the decline, and it may even speed it up. Laws can be effective, but only to the extent that they are accepted and supported by those who are bound by them. Judges cannot order wells to stop declining; all they can do is order the owners of the wells to stop pumping. That will work only if the owners have other resources available. If they do not, they will either disobey the law or change it.

In Colby, Wayne Bossert talked about Groundwater Management District #4's failed attempt to regulate its way to zero depletion. "It was a very sophisticated proposal," he told me. "It was very regulatory. It took our board a long time to get their arms around it, and I'm certain the people that we presented it to didn't understand it. There were a lot of really super ideas in there, and I expect that in some form or another a lot of those ideas are going to translate into something. They actually did translate into a state proposal, which has suffered the same fate as ours."

In Bossert's mind though, the fate of the zero depletion proposal—either at the district level or at the state level—did not matter as much as what came afterward. "A bunch of us met once every three weeks for seven months, and we hammered out an agreement. There were no state agencies or eastern Kansas people involved; it was a bunch of local landowners and water users and people involved in groundwater management, sitting down, and we came up with this plan: We are going to identify high-, medium-, and low-priority aquifer subunits. The high-priority subunits are going to have enhanced management; they're going to set goals and implementation schedules to meet those goals. The overall goal is to 'improve the decline situation.' We wrote that so the individual districts can decide what it means. The mediums and the lows can go on as they are a little bit longer, until they become a high-priority area. Let them go the way they're going to go." He leaned forward. "That doesn't mean 'do nothing.' You're always going to promote conservation and efficiency. We're going to hold their feet to the fire to use the water well. But I'm not going to make them capitalize up to do it. We're going to take the systems they have and help them use those as well as they can."

As long as "as well as they can" truly means as well as they *can*, that is a very difficult policy to argue with.

In Great Bend, Kansas, over the long eastern limb of the Ogallala, I heard a tale of two wildlife refuges—a pair of major wetland complexes called Quivira and Cheyenne Bottoms that feather out each spring and fall as masses of birds flock along the great migration route ornithologists call the Central Flyway. Quivira is a federal wildlife refuge; Cheyenne Bottoms is protected jointly by the state of Kansas and the Nature Conservancy. Both refuges hold senior rights granting them the use of surface water, Quivira from Rattlesnake Creek, Cheyenne Bottoms from Wet Walnut Creek and the Arkansas River. There is little surface water to use. Center pivots have sprouted throughout the watersheds of all three streams, and the baseflow in each has dropped dramatically. In the Arkansas River it has gone away entirely. Baseflow in the Arkansas officially begins at Great Bend; above, there is nothing in the riverbed except ephemeral runoff.

In the early 1980s, environmentalists launched an administrative appeal through the Kansas Water Board to restore water to Cheyenne Bottoms by cutting back on groundwater irrigation in the basin of Wet Walnut Creek. The appeal was successful; the water began to flow again. The environmentalists were pleased. The farmers were not, especially those who had been served with orders to stop their center pivots. Countersuits were threatened. The classic social fracture lines, "elite birdwatchers" versus "land rapers," appeared and began to calcify.

At Quivira, refuge superintendent Dave Hilley watched the dispute over Cheyenne Bottoms and wondered if there might not be a better way. "We were going to get the water," Hilley told me five years later. "We could always go to court and get it; there are only three rights in this groundwater district that are senior to ours. So we had that hanging over the situation. But we saw what had happened at Cheyenne Bottoms, and we didn't want to go that way. There were a lot of hard feelings. The environmental community may have won legally, but they lost politically. We decided to try a different tack."

Hilley made his pitch at a crowded meeting in the basement of a bank in the little town of Stafford, the location of the headquarters for both the wildlife refuge and the Great Bend Groundwater Management District. "I told the people who were there that we should work together. We all need the water, so let's all try to take care of it." A committee was formed. Warily, the irrigators and the birdwatchers began to labor side by side to create a plan to share the water in Rattlesnake Creek in an equitable manner.

The Rattlesnake Creek Management Program Proposal was released on June 29, 2000, under the authorship of the committee, now called the Rattlesnake Creek–Quivira Partnership. The proposal called for the outright purchase and retirement of some of the water rights in the creek's drainage basin on a willing buyer–willing seller basis. It called for the creation of a water-rights bank in which irrigators could deposit their rights and withdraw them again as needed, without fear of losing them through inactivity. It called for averaging the water use required to maintain a right in good standing over a five-year period, instead of just one year, to allow farmers the flexibility to voluntarily reduce their water use in times of drought. The irrigators gave up their end guns, attachments on the ends of center pivots designed to spray water into the corners of square fields; the refuge gave up several miles of perimeter ditches, reducing the amount of water required to maintain the wetlands. Weather stations would be installed to provide the data required for irrigation scheduling. Enforcement would be stepped up.

Part of the proposal required changes in state law and in federal wildlife policy, and that allowed the partnership to demonstrate its unique makeup to the rest of the world. "We sent a delegation to Washington consisting of a guy from the Audubon Society and a guy from the Farm Bureau," Hilley smiles. "They got there and I started getting calls from wildlife service headquarters. 'What the heck's going on out there? We've never seen these guys work together before.'" With traditional foes united behind the proposal changes, they passed easily. The program moved from proposal to implementation.

As I write this in late 2004 it is still too early to tell whether the Rattlesnake Creek Management Program will save both Quivira and the farms; if it does not, it is doubtful that anything else can. At the least, it will ease the transition to the Ogallala's end in this region. That transition has been feared for a long time. It has been speculated about, worried about, and argued about. Now, as it approaches, the speculation, worry, and argument seem almost irrelevant. We can, and should, prepare. But the most important preparation may consist of accepting the inevitable and getting ready to move on.

ONE AGRICULTURAL TECHNIQUE that may make a large difference in how well we are able to move to a post-Ogallala economy is what is generally called no-till farming. In contrast to traditional methods, where crop stubble—the leaves, stalks, and other unused parts of the crop plants—is either turned under the soil or burned shortly after harvest, no-till leaves the stubble in place in the field until planting time. The field looks messy, which causes some farmers to approach the technique gingerly. But the ground stays moist.

"Where the savings come is in residue management," explains agronomist David Nielsen in Akron, Colorado. "As you get more and more residue, you reduce the movement of water. If you keep it layered close to the soil, evaporation slows down."

Asked if no-till allows dryland farming for crops that might otherwise need irrigation, Nielsen responds with a cautionary yes. "Everything is lower-producing in the dryland situation," he warns. "But, yes. Corn production had left this area, and now it's back, in a big way. It's a large part of dryland cropping systems out here today, because no-till is saving more water."

Across the table, Wayne Shawcroft points out that the water saved through no-till allows more intensive crop rotations. "The old dryland system was simply wheat-fallow," he explains. "You fallowed fourteen months to store enough water to grow a crop in the subsequent ten months. Your fallow efficiencies were pretty low; if you were saving 25 to 30 percent of the moisture, it was pretty good. As techniques improved, we were starting to save, experimentally, as much as 45 percent of the fallow-period precipitation. If you do that, and then you have normal growing-season precipitation, you have more water stored than the crop can use. So people started adapting to that, and dryland corn and other rotations started coming in. Now it's quite a bit more common to do a fallow, a wheat, a corn, and then fallow again. Or millet, or sunflower—several different possibilities. That's changed over—what? The last fifteen or twenty years?"

"It's mostly the last fifteen," says Nielsen.

"When I first came here, in the 1950s," Shawcroft continues, looking a little dreamy, "you used to see a field of dryland corn once in a while, but if they got fifteen bushels an acre they were lucky. Now it isn't uncommon to raise a hundred and twenty bushels on an acre of dryland corn with this cropping system. On sixteen inches of rainfall. So it's changed totally." He smiles. "Some people look at having to go back to dryland as just a disaster. Well, Washington County here, where Akron is located, is still basically a dryland county. And looking at it from a bank standpoint, which I do these days, Washington County's dryland farmers are much better off than a lot of the irrigators. People who have adapted to this, who are raising seventy-five to one hundred bushels of dryland corn, compared with trying to aim for two hundred bushels with irrigation—you give me these dryland people any day."

"There certainly was a lot of pride in a clean-plowed field," remarks Nielsen. "How beautiful that was. The residues didn't fit that image for a long time. But now . . ."

"Now it's the other way," says Shawcroft.

"And a field that is not weedy, that has all this *pretty* residue—a lot of people have an appreciation for that now who didn't, twenty years ago."

"Was it a hard sell?" I ask.

"Oh, yeah," says Shawcroft.

"You still have a mentality out there that no-till farming is trash farming," agrees Nielsen.

"But now it's commonplace," says Shawcroft. "You look at somebody who's doing the old wheat-fallow system as a rarity. There's just a handful of people who are actually doing it."

"Environmentally," Nielsen observes, "we might have seen a dust bowl here in the past few years if we hadn't been using no-till methods. But we haven't had that disaster."

"Dryland's not a bad word, with the management now."

"And when you talk about what's going happen to overall productivity as the Ogallala runs out, people are always afraid it will drop. But we've had a general *increase* in acres farmed here, because we went from one crop in two years to two crops in three years, or three crops in four years. That higher percentage of cropping in dryland has increased productivity over the whole area."

YOU WILL NOTE that Nielsen mentioned a "dust bowl." That is a recurrent fear of those who find themselves forced toward dryland farming. If the rains do not come, will the soil, even soil cultivated with no-till, stay in place without putting water on it? The "black blizzards" of the 1930s, when dust blew so thickly that cars wouldn't run and people and animals caught outside died from suffocation, still cast long shadows. The stories are endlessly repeated; the photographs of dust-blocked doorways, decrepit automobiles, and lean, haunted people stalk our national psyche. The farms of the Dust Bowl were not irrigated. If we stop irrigating, do we risk its return?

A thing often forgotten when this question is being considered is that the "Dirty Thirties" were not an anomaly. They were followed twenty years later by the "Filthy Fifties." The Palmer Drought Severity Index, a statistical tool used by meteorologists to measure dry spells, bottomed out again over the midsection of the continent and stayed that way for six years; at its nadir, in 1954, the afflicted area covered a swath reaching from Nevada and Idaho south into Texas and east all the way into Virginia. Soil was blowing over forty million acres. Why are there no traveling exhibits or picture books to memorialize this equally terrible time?

There are a number of possible answers to that question, but they all sift down to this: It was not equally terrible. The drought was as bad but the damage was much reduced. Some of that can be attributed to better cash reserves (the Dust Bowl years corresponded, incidentally but with ominous precision, to the Great Depression), but most of it was due to improved management. Lessons learned in the 1930s were put to use. Windbreaks and shelterbelts had been planted around most homes and along the edges of many fields. Plowing was done crossways to the prevailing wind, to create speed-reducing turbulence at the ground surface. Fallow fields were deep-plowed but not disked, putting large clods on the surface in place of crumbled soil. These and other techniques, including a general increase in irrigation, kept damage below the national panic threshhold. In April 1935, as the Senate was debating a bill to provide relief to drought-stricken areas in Kansas and Oklahoma, a dust storm darkened the skies above Washington, D.C., and parts of Kansas and Oklahoma began passing over the senators' heads. Nothing like that happened in the 1950s. Nothing like it is likely to happen as dryland farming returns.

There are, however, no guarantees. Drought will certainly recur, and this time there will be no underground ocean to bail us out. That could make a very large difference.

Not all of the lessons from the Dust Bowl have survived. The circular plowing required by LEPA sprinklers provides only half the wind resistance of a cross-wind plowed field, so it is only half as likely to keep its soil in dry, windy conditions. Shelterbelts have largely disappeared as farmers have stretched their center pivots to the edges of roads to squeeze all possible production from each quarter-section. These changes may seem foolish when the rains dry up and the wind rises. Or they may not. The standing stubble left by no-till agriculture may take care of the problem. The buffalo may come back. It is too early to tell. The one thing certain is that irrigation will fade and that dryland will return.

Ray Brady's early impressions of dryland were formed during the Filthy Fifties. "My mother's family lived in Lamesa, in the southern part of the panhandle," he reminisces, looking out over the flat green fields of the Llano Estacado. "I can remember coming from south Texas in the summer, going from Big Spring to Lamesa at maybe thirty miles an hour with the headlights on and dirt blowing across the road so you could barely see. Going that whole way in whatever old car we had with the windows up. That was before air conditioning, and it was *hot*. So I think farming methods have advanced, in that respect. I'm not sure I agree with plowing all that country up. I've got a photograph of my grandfather and his guys standing around the ranch in Zavala County. You look at the grass, and then you look at what little grass is there now, and you wonder where it all went. But that's what that country all looked like. And now it's all cultivated. Did we do the right thing? I don't know. What is the right thing? What do you do with all the people? And I guess what we have to consider is whether it was or wasn't the right thing, that's what did happen. So how do we go forward?

"I think that's how our groundwater district is trying to look at things. This is the situation: Now what do we do? Do we try to regulate everybody out of business? Is that a feasible alternative? It's certainly an alternative, but is it practical? How do you manage the transition? At some point, things become uneconomical. And I guess you could set that uneconomical level politically. You could say, 'Okay, depress the price for wheat and grain and corn.' That would solve something here, but it might also lead to more cotton farming. So then you go fiddle with cotton, and pretty soon you're fiddling and fiddling, and you've got a web of inconsistencies that doesn't work. I guess we could debate foreign policy, too, and go off on another one. But all this comes down to where is the water, and where is this going?"

"What's the country going to look like here in fifty years?" I ask.

"Good question," says Brady. "I don't know."

Epilogue

THE BUFFALO SPRING

IT IS A DAY OF GREEN GRASS and little winds, topped by an Ogallala blue sky with a thin white marbling of clouds, and we are bouncing over fire trails on South Dakota's Pine Ridge Indian Reservation in a four-wheel-drive Chevrolet Suburban belonging to Trudy Ecoffey's mother. Outcrops of fine-grained Arikaree sandstone, known locally as "butte rock," accent the ridge crests. Northward lies the White River Valley and beyond that, the rolling Missouri Plateau; southward, the High Plains stretch away to Texas. Somewhere out there are Trudy's buffalo.

Pine Ridge is where the High Plains stop. It is the northern equivalent of the Caprock Escarpment in Texas or the Mescalero Escarpment in New Mexico. A thousand feet tall and as much as twenty miles broad, it meanders for more than one hundred miles through northern Nebraska and southern South Dakota. Mazes of side ridges finger out to the north; small streams fed by Ogallala springs tumble through the valleys between them. The scent of Ponderosa pine hangs sweet and resinous on the afternoon breeze.

The land we are traveling over is near the eastern edge of Shannon County, which is wholly on the Pine Ridge Reservation. Shannon has the unsought distinction of being the poorest county in the United

States. More than half the population lives below poverty level; when
only families with small children are counted, that figure climbs to
more than two-thirds. Per capita income is just a shade over six thousand dollars per year. There are no significant industries. Not even
tourism has made much headway here; though it is one of the more
attractive parts of South Dakota, containing the southern section of
Badlands National Park as well as the Pine Ridge itself, the county has
not been able to capitalize on its visual beauty. Twenty-seven times as
many visitors enter the national park from the north, through Pen-
nington County, as enter it from the south. Nearby state parks and
federal recreation lands, all of them in Nebraska, siphon off hikers,
photographers, and natural history buffs. Shannon is possibly the
only county in the United States in which there are no motels.

For seventy-one days in 1973, Shannon County was the site of the
last Indian war (so far) in America. On February 27 of that year, a
mixed crowd of three hundred Lakotas and a smaller number of
American Indian Movement activists gathered at the tiny village of
Wounded Knee—the site of an infamous 1890 massacre of Lakota
"ghost dance" practitioners—to air a list of grievances. The airing
never took place. Finding themselves surrounded by FBI agents and
National Guard troops, the Lakota and the activists threw away their
list and declared the Pine Ridge Reservation an independent nation.
The federal response was to block all roads into Wounded Knee, cut
off the hamlet's electricity, and demand that the leaders of the insur-
gency be handed over. The Lakota painted themselves in the manner
of their ancestors and refused. There were daily exchanges of gunfire.
Three weeks into the war, the feds declared an amnesty and opened
the roadblocks, expecting the occupiers to leave; instead, more pro-
testers arrived, swelling the forces at Wounded Knee to more than one
thousand. The roadblocks were reclosed and the siege was rekindled.
All told, more than 130,000 rounds of ammunition were fired into

Wounded Knee—about eighteen hundred rounds per day. Two Lakota were killed.

Negotiations ended the Last Indian War on May 5, 1973, with promises of amnesty and an agreement by the government to address the Lakota's grievances. Like most promises to Native Americans, these were not honored. Twelve hundred people were arrested; many who were not arrested complained of official harrassment for a long time afterward. Sixty-four unsolved murders took place on the reservation over the next three years. Leonard Peltier, an AIM leader, was arrested in 1976 and charged with killing two federal agents; he remains in prison today, convicted of crimes that many believe he did not commit. The issues continue to simmer. On January 16, 2000, a small group of activists took over the Red Cloud tribal building in the town of Pine Ridge, occupying it for eighteen months. Having, perhaps, learned something from the events of 1973, authorities simply ignored the Red Cloud occupation, and it eventually faded away.

Trudy Ecoffey lives in a yellow house trailer four miles north of Wounded Knee, with a husband named Lee and a small daughter named Echo Dawn and a 115-pound German shepherd named Ash. A tall, slender young woman with a wealth of auburn hair and a master's degree in range management from the University of Nebraska, she is a member of the agriculture-business department at Oglala Lakota College, where she has developed a widely admired course in bison science utilizing the wisdom of tribal elders as well as textbook range ecology. Part of the course involves hands-on experience with a free-roaming herd on the Knife Chief Bison Range a dozen miles or so south of campus, and it was in the course of this work that she began to notice the springs. Knife Chief had cattle on it until only a few years ago. While the cattle ran there, most of the pasture springs dried up. When the buffalo came, the springs started flowing again.

There are four of us in the Suburban, which Trudy is thinking about buying from her mother. Trudy is driving; I am in the shotgun position in the front passenger seat. My wife is directly behind me, on the right end of the middle seat. Beside her, on the left end, perches biologist Jim Taulman. Jim is also on the faculty at Oglala Lakota College, and his presence today is driven by something more than simple curiosity. He is trying to develop a remote-sensing model for vegetation mapping on the Pine Ridge, and this outing will give him a chance to do some ground truthing. He has heard from Trudy about the resurgent springs, but he has not yet seen any of them.

"My buffalo didn't come in yesterday," Trudy had warned us as we started out. "They're supposed to come in today, so I have to somehow get over there." That determined which spring we would visit. The new animals that are to be delivered—four pregnant cows—will join a herd of perhaps twenty-five already roaming the two-square-mile Knife Chief range under the care of bison manager Ed Iron Cloud, whose family once lived on the land the range now occupies. The area where they are to be released is near a spring. It is not one of the resurgent springs—oral tradition suggests it has never stopped flowing—but it isn't on any of Jim's maps, either. It will give him a new point to check on the satellite images, and it will show me what the buffalo-pasture springs look like.

Heading south up Medicine Root Creek from the town of Kyle, we quickly got into a maze of tiny dirt tracks, built primarily as fire trails, that crisscrossed the long slant of the Pine Ridge. The Suburban traversed hillsides, climbed through gaps, crossed what might have been creeks in damp weather. Several times I had to get out to open gates. Once we passed a small tumbledown home buried back in a grassy valley. Two decrepit cars sat in the yard, laundry hung from a line, and a small boy looked up from his play as we passed. Mostly the road was two bare tracks, one tire width each, through foot-high grass. Occasionally it disappeared altogether and Trudy seemed to be navigating by bent grass stems. Finally we curved over a tall ridge and came down—I could not tell you in what direction—into a steep green valley with a copse of Ponderosas partway down its slanting floor. On a little shelf on the far side of the valley were the remains of a cabin: a roof slanting into the ground over piles of disintegrating logs that had once been walls. A full stock tank stood to the ruined building's left. A track even more obscure than the one we had been following led toward it.

"That's a nice old cabin over there," I remarked.

"Yup," Trudy agreed. "That's the old Iron Cloud homestead. You can see the well, and—oh, there's the buffalo. By golly."

On the slope behind the cabin, beneath a ragged row of Ponderosas ranged like tired sentinels along the crest of the ridge, a line of twenty or so large, shaggy brown creatures cropped casually at the short grass. These were the resident bison, the herd that Trudy expected to increase by four before the day was out.

"There they are," I breathed.

Trudy spun the wheel. The Suburban obediently turned onto the track leading to the cabin. We crept toward the animals, who failed to spook; on the contrary, they seemed to be sidling closer.

"Wow," said my wife softly.

I was fumbling with my camera, babbling, "What a deal. What a deal." The telephoto's threads wouldn't catch. I got it on the fourth try, looked up, and immediately felt foolish. The buffalo were all around us. It would have been an easy shot with the standard lens; with the telephoto, I could fill the digital camera's viewscreen with part of one huge, shaggy head.

Trudy had the driver's door open and was standing on the sill, a camcorder held to her eye. "Any closer," she remarked, "and I'm on top of the vehicle. These are wild animals, and they're pretty unpredictable. You may have to drive us out of here."

"What'll your mother do if a buffalo gores this thing?"

"I'd prefer not to worry about the *car* just now."

"What do you think is attracting them?" asked Jim.

"They want cake. We give them supplemental feed out of pickups, so they've learned to associate vehicles with food. It's a good thing I don't have the tailgate down—we might have one of those big ol' heads right in here with us."

Eventually the buffalo seemed to conclude that we weren't a promising source of handouts. They wandered off slowly in three directions. One calf clambered over the wall of the stock tank and into the water. Trying to climb out again, it got hung up on its chest, eyes frantic, front feet flailing the air.

"We'd better not try to help him," said Trudy, lowering herself back into the driver's seat. "Momma might misjudge our intentions." She backed the Suburban around. "By the way," she added, "this is where the spring is, in this ravine. We'll move on around to where it comes out, and we can walk down."

A few minutes later she was parking the big vehicle atop butte-rock pavement on the far edge of the ravine. Ponderosas surrounded us and spilled down the side of the little valley. A trail moseyed down-slope. Trudy strode onto it; I followed, keeping one eye peeled for buffalo. It was less than an eighth of a mile to the spring, trickling out of Arikaree sandstone in the ravine bottom, surrounded by a barbed-wire fence. Where it ran under the fence and on down the ravine the buffalo had muddied it, but within the enclosure the water looked clean and fresh. Duckweed swam on its surface.

"This one isn't new?" I asked.

"No," Trudy responded. "It's been here a long time, ever since they first came out here and started homesteading. As far as we know, it's never stopped running."

"Turkey," said Jim. A pair of wild turkeys headed for the spring shied away when they saw us. They moved through the underbrush on the far side of the ravine, two large shadows with slender, jerking heads.

"But there are new springs like this coming in?" I asked.

"Yup," said Trudy. "They're popping up all over the place. There's some over around the town of Pine Ridge that came up last year, and that was a drought year. The whole underground hydrology system around here is just . . . strange. Nobody knows what's going on."

"Any place other than buffalo pastures?"

"Not that I'm aware of."

"Do you have a clue as to why?"

She shook her head. "We really don't. It might have to do with the time buffalo spend at their water source. Cattle will spend hours hanging around a spring; they just stomp it down. The bison come down and drink out of it, but they won't stay. It's just their natural instinct. They don't wait around at their water source, because that's also where the wolves would be, or the mountain lion. So they don't end up stomping it out. But I talked to a riparian scientist, and she said that even a once-a-week visit from bison-sized animals would cause damage to a place like this."

"I see you've got it fenced off," I pointed out. "When did you do that?"

"Oh, this was done three or four years ago, I think," said Trudy. "I don't know if it was for the buffalo—probably for the cattle."

Jim perked up. "Cattle are run here?" he asked.

"Cattle *were* run here," said Trudy. "That's why we wanted to study this area. Cattle previously grazed it, and now bison are grazing it, and we want to document what kind of changes are happening. There are certain things that have changed already. I'm not trying to bad-mouth cattle; I just see bison as a different animal. We'd like to know what they do that makes them different."

"So you think maybe its mechanical?" I asked. "The springs are coming back because the cattle aren't trampling them anymore?"

"Well, that and good grass management," said Trudy. "Here on Knife Chief, the grass, even after this drought year, is exceptional. I don't know if that has that much to do with the bison yet, because we've only had bison on it for a year. But the soil holds the water, because all the grass is in place. It's not running down the hillside—it's soaking into the ground."

I had a brief, fleeting vision of restoring the Ogallala by bringing the buffalo back. Perhaps Frank and Deborah Popper were right. There were nearly twice as many buffalo on the High Plains in the early nineteenth century as there are cattle today. The range was undeniably in much better shape then than it is now. Bring back the buffalo, and you bring back the grasslands; bring back the grasslands, and you bring back the springs. Restoration through livestock choice. It wouldn't work. There is, to begin with, the problem of fences. The historic buffalo range didn't have them; the great herds ran free over hundreds of thousands of square miles of open prairie. They never stayed in one place long enough to cause lasting damage. That couldn't happen today. It is actually fences, as much as cattle, that cause what we think of as "cattle damage." If the cows could roam as the buffalo did, would they choose to stay in one place and eat it down to the bare, packed earth?

But even if the Popper's dream of a great, fenceless Buffalo Commons where bison herds of different ownerships could mingle and move as their hairy ancestors did came true, it wouldn't bring back the water. Not where the problem is overdraft. And overdraft is the problem almost everywhere. Buffalo will not bring back the springs of the Llano Estacado, not while the irrigation pumps run and the Rule of Capture is king. Not while urban Texans demand lawns the same shade of green as New England's. On the southern High Plains, much Ogallala water currently goes to dairies; I wouldn't want to be the guy in charge of milking a buffalo herd. In Kansas and New Mexico there are fishing lakes in state parks that are kept full by pumping water from a hundred feet and more underground. How are buffalo supposed to help with that?

But buffalo are certainly part of the answer. In a place like the Pine

Ridge, they may be most of it. If we can bring back the springs by changing the grazing pressure on them—not reducing it, simply changing it—then by all means let us do that. There is no one-size-fits-all solution. In Kansas, the solution may be subsurface drip irrigation and carefully honed agreements among multiple user groups. In Texas, it is certainly a change in the Rule of Capture. Here, it is probably buffalo.

WE ARE BACK IN TRUDY'S mother's Suburban, back on the fire trails, driving—I think—north. Trudy and Jim are engaged in a dialog about the difficulties of doing research at a small college, with a tiny budget, a large class load, and an administrative staff that hasn't really figured out what research is yet. The Suburban crests a high, grassy ridge. Jim is in the middle of a discourse about how he must bring in equipment he has purchased for personal use to teach his field courses when Trudy suddenly interrupts him: "There come the buffalo, right there!"

On another green ridgetop, a half-mile or more distant, a red pickup pulling a livestock trailer is moving slowly against the blue sky. There is there and here is here, and in between is a gulf like a grassy Grand Canyon.

"That's your buffalo?" asks Jim.

"Must be," says Trudy. She spins the wheel to the left. The Suburban obediently leaves the fire trail and starts down into the Grand Canyon.

I grab the seat involuntarily. "You'll get this thing totally stuck."

Trudy laughs. "I hope not." The big vehicle moves surefootedly downward, following a side ridge that is slightly less steep than the rest of the canyon wall. Around us a slanted sea of short grass wears its April green, interrupted occasionally by small ledges of butte rock. We leave the first ridge and start over a second. "I'm a little bit familiar with this," Trudy reassures us, as the red pickup and its trailer appear briefly on the skyline and then disappear again. "I wonder who's leading him out here? He said he didn't know where he was going. I wonder where Ed Iron Cloud's at?" The pickup comes into view again, looking somewhat closer. "He's coming down this way," she says. "He's on a different trail. Well, I guess it doesn't matter which way we go. Now where's he going?"

"He's turning off."

"He's turning around."

"He stopped."

"It's probably good not to have them dropped with the other buffalo," remarks Jim.

"Right away, anyway," agrees Trudy as the pickup disappears once more. "I wonder how he got in here? He told me last night he wouldn't know how unless Ed showed him the way."

"Maybe Ed's riding with him," I suggest.

"That could be," says Trudy. "That very well could be. I didn't think about that." She drops the Suburban expertly over another small ledge of butte rock. We reach the muddy bottom of the canyon and turn upslope toward our last visual fix on the pickup, grinding smoothly upward over the same type of grass-and-butte-rock terrain we just descended. In the backseat, Jim chortles.

"This is a nice truck," he says. "Pay your mom the money for it."

A few minutes later we pull up beside the red pickup, on the passenger's side. Sounds of heavy movement come from the livestock trailer. Trudy rolls down her window.

"Have you been out stealing buffalo again?" she calls across to the man in the passenger seat.

The man grins and gets out of the pickup, walking bent at the waist, as if permanently set at ten after six. Wire-frame glasses rest on a prominent Lakota nose; long black hair, shot through with gray, is pulled back in a ponytail. This is Ed Iron Cloud, the driving force behind the Knife Chief Bison Range. Until recently, Knife Chief was just another reservation cattle allotment held by a white rancher, a holdover from Bureau of Indian Affairs policies of the early twentieth century. These leased allotments have always been a thorn in the side of reservation residents, but the leases have proved difficult to break. Ed Iron Cloud found a way. There is, in Lakota culture, a social unit called a *tiospaye*. Roughly translated, the word denotes an extended family, or clan. But it also means "community." It has ties to a specific locale, and in presettlement days, elders from the various tiospaye formed the ruling circle of the tribe. Iron Cloud found a clause in the Oglala Lakota constitution that specified that reservation communities had priority for grazing allotments on reservation land, and he argued successfully to the Bureau of Indian Affairs that the word "community" should include the tiospaye. The white rancher and his cattle were evicted, and the Knife Chief Tiospaye, to which Iron Cloud belongs, took over the lease and began running buffalo. Some of the animals belong to the tiospaye; some belong to the college. The college is not charged board for its buffalo and is allowed full use of the Knife Chief range for research and teaching purposes. In return, the tiospaye claims ownership of any calves born to the herd.

"We've got research plots," Trudy had explained earlier. "Hundred-by-hundred-foot exclosures that the bison can't get into. We're doing grass sampling and soil sampling. As far as we know, there's never been any soil data collected over a long period of time in bison pastures. Soil doesn't change that quickly, so it's going to have to be a long-term study, but we have to start somewhere." Most of the upper foot or so of grassland soil is made up of humus—organic litter, plus living and dead microorganisms and their waste materials. Along with other range ecologists, Trudy has theorized that the humus community found under bison range has a different species composition than that found under cattle range. The differences, and the effects those differences have on such matters as water retention and aquifer recharge, may be trivial or profound. No one knows.

"Maybe it'll change, and maybe it won't, because bison are out here," observes Trudy. "We don't know. But it seems like it's time we checked."

We all pile out of the Suburban to join Iron Cloud and the pickup's driver, a young Native American man in blue jeans and a blue denim jacket. Without further preamble, the driver walks to the back of the livestock trailer and lowers the ramp. He grasps the gate latch.

Jim is standing to one side with his camera. "Is this a good place?" he asks.

"How fast can you run?" responds the driver. He opens the gate. A hairy earthquake of bison flesh erupts down the ramp, resolving itself into four fleeing animals. Shying away from us, they gallop off a hundred yards or so and pause, glaring back suspiciously, a primeval vision of brown buffalo, green grass, and blue sky. The driver looks over at Jim.

"Actually," he says, "after a ride like that, they want to get as far away from anything human as they can. As long as you're not directly in their path, you're fine."

For a long, timeless moment four buffalo and six humans regard each other across a gulf of space and a greater gulf of species. Small winds move in the sunshine. A meadowlark's song, uncontrollably optimistic, floats in from some secret place beyond the sky.

Eventually, one cow appears to make up her mind. She heads for the ridge behind us, making a wide circle below us to the left. The other three animals follow, trotting nervously, their eyes as much on us as on the way ahead. The little herd gathers speed as it passes. It thunders over the green rim and disappears, four sets of narrow brown hindquarters and kicked-up heels carrying the hope of the herd, the Knife Chief Tiospaye, the Lakota people, and perhaps the Ogallala Aquifer with them.

ACKNOWLEDGMENTS

Traveling for many weeks across the flattest landscape in North America could have seemed downright boring, and I have many people to thank for the fact that it did not. Foremost among those is my wife and traveling companion of many years, Melody James Ashworth, who not only endured all that time with me crammed into the tight quarters of a Toyota Prius and a series of not-always-particularly-great motels, but who made it all seem easy and pleasant—and whose insights, as a trained biologist, into the landscape we were traveling over were usually far deeper and more lucid than mine.

Five other people made outstanding contributions to this book, and it is probably no accident that all of them are scientists and that three of them are geologists. Ray Brady, Jim Goeke, and Jon Mason not only provided solid grounding in the structure and hydrology of the Ogalalla Aquifer, but proved thoughtful conversationalists and excellent companions in the field as well. (Ray also—not incidentally—has memorized the coordinates of all of the best sparerib houses in North Texas.) Mahbub Alam may know more about the uses of groundwater after it has been pumped out of the ground than any other person in North America. And Trudy Ecoffey's work as a range ecologist on Native American lands has given her a unique perspective on the shape of the future of the High Plains after the easy water goes away—not to mention a truly extraordinary ability to maneuver a four-wheel-drive vehicle over landscapes most of us would have had trouble negotiating on foot.

Beyond these six, singling out any one person's contribution gets difficult, and it seems best to simply go state-by-state.

In New Mexico, Roy Cruz gave me an excellent overview of both the geology and the politics of the southern Ogallala, as well as providing the information that led me to Dennis Holmberg and the Lea County Water Users Association. Dennis, in turn, made up for missing our first appointment by showing up for the second with his wife on the morning of their wedding anniversary and sharing his view of declining water tables as both a water users association officer and as a county official. (I should also thank Debra Holmberg for her grace and good nature for putting up with that intrusion on their special day.) Dennis's secretary, Lue Ehrlich, not only found Dennis when he had left the office early; she also found a pinch hitter, Bob Carter (also of the Lea County Water Users Association), who gave us much time and information on extremely short notice. Jim Whary, who is helplessly watching his cottonwoods die at Oasis State Park, also provided much information on short notice—in his case at the end of the work day when he should have been going home. Blackwater Draw curator Joanne Dickenson gave us a personal tour of the oldest well in North America; her knowledge of the ancient peoples she studies is encyclopedic, and her enthusiasm for them is contagious. And we couldn't have reached Joanne without the help of Robin Gillispie, Tori Myers, and Ben Aubuchon, the chain of archaeology students at Eastern New Mexico University—Ben to Robin to Tori to Joanne—who put us in touch with her.

In Texas—along with the massive contribution of Ray Brady—there were county agents Dana Porter in Lubbock and Leon New in Amarillo, who shared their experiences helping farmers continue to grow crops above a decreasing water table. (Leon also gave us a history lesson in the development of LEPA, or perhaps, as one of the originators of that system, he *was* the history lesson.) Jeff Johnson at Texas Tech University provided useful perspectives on declining water tables from the field of agricultural economics, and Chris Wingert at the Colorado River Municipal Water District in Big Spring was an able defender of weather modification as a cost-effective means of increasing water supplies. Boone Pickens generously took time from an incredibly busy schedule to provide an extended interview about Mesa Water (thanks to Sally Geymuller for setting that up). Eddie and Patti Guffee of the Museum of the Llano Estacado in Plainview provided plentiful information on the history of water development in the Texas Panhandle. And Seyf Ehdaie—formerly the charge d'affaires at the Iranian embassy in London, now a professor of environmental science at the University of Texas of the Permian Basin in Odessa—entertained us royally with stories of life at the extreme southern end of the Ogallala over an excellent lunch at Harrigan's.

We did no interviews in Oklahoma; but James Horne of the Kerr Institute in Poteau provided background data beforehand, and Ron Bell of the U.S. Army Corps of Engineers office in Tulsa was generous afterward with information about Optima Dam.

In Kansas, the list is a long one. Dave Hilley of the Quivira National Wildlife Refuge and Dan Zehr of Groundwater Management District No. 5, both in Stafford, explained the Rattlesnake Creek–Quivira Partnership in depth and with much enthusiasm. Hank Hansen in Garden City and Wayne Bossert in Colby brought many years of experience directing groundwater management districts to conversations that were informed, lucid, and (especially in Wayne's case) philosophically deep. Hank also arranged a meeting in his office that brought together himself, Mahbub Alam, farmer and water attorney Mike Ramsey, and Dave Brenn, former manager of the Garden City Corporation—the largest agricultural operation in southwest Kansas—for a wide-ranging discussion of water issues on the High Plains. Dave Wehkamp, near Ingalls, provided a farmer's perspective on irrigation choices, and Larry Stange in Garden City gave us useful insights into the economics of sprinkler systems from his position as an irrigation equipment sales representative. And of course there was Mahbub, who not only shared his own deep pool of knowledge of irrigation systems but introduced us to Dave and Larry, provided transportation to visit them, and patiently answered all our questions about their work, and his own, afterward.

Our time in Nebraska was dominated by the giant figure of Jim Goeke. Others, however, were important as well. Jeff Beuttner of the Central Nebraska Public Power and Irrigation District in Holdrege gave us much information about the Republican River controversy, and about his agency's role in the river's fate. Dean Yonts and Steve Sibray of the University of Nebraska's extension office in Scottsbluff explored the intimate (and often overlooked) connection between surface water and groundwater from a very practical perspective. Clint Carney and Jim Kramer of COHYST made room for us at their drilling site on John Jensen's ranch, and shared their enthusiasm for the new form of geology that they practice, which involves computers and GPS systems as much as it does fieldwork and laboratory analysis. Tom Downey's enthusiasm for his drill rig was infectious, and his two crew members, Chris Howard and Aaron Withington, handled the complex machinery of the rig with skill and precision. Television journalist David Fudge, who was taping the drill site for station KNOP in North Platte, graciously shared his raw interview videotape. And John Jensen was not only a friendly host but an eager onlooker, as enthralled as the rest of us by what was coming out of the earth beneath his ranch.

In Colorado, agronomist David Nielsen of the Central Great Plains Research Station in Akron set up—and participated in—a fertile conversation with geohydrologist Joel Schneecloth and banker (and exirrigation specialist) Wayne Shawcroft; and our old friends Jim and Margaret Frye, of Fort Collins, provided some much-needed R&R in the middle of an otherwise intensive five-week journey. (Jim and his son Dan also located and photocopied several hard-to-find research papers from the Colorado State University library—including Jim Goeke's master's thesis.)

In Wyoming, Jon Mason not only devoted much of a working day to giving us a guided tour of the Gangplank and shared his contagious curiosity about all things underground, but introduced us to John Harju in the Office of the State Engineer, who gave us a quick but thorough grounding in Wyoming water policy and water law.

And in South Dakota, Delano Featherman and Ed Iron Cloud of the Oglala Lakota tribe shared the current perspective on the Ogallala of the people the aquifer is named after. Jim Taulman of Oglala Lakota College not only introduced us to Delano and—memorably—to Trudy Ecoffey, but came along on both interviews and asked some of the pointed questions I should have come up with myself.

Three other people should be thanked by name here. Warren Wood of the U.S. Geological Survey in Reston, Virginia, was not able to meet with me, but he did send along a pair of papers he had written on playa lakes on the Llano Estacado. David Tarkalson of the University of Nebraska Research and Extension Service facility in North Platte filled in valiantly for Jim Goeke for a brief interview while Goeke was out of town. And Steve Fox of State Farm Insurance sliced speedily through all the red tape and got our claim properly filed when the winds of Kansas quite literally tore the front passenger-side door part-way off the Prius just outside of Garden City.

Numerous people whose names were never known to me, or have been forgotten, contributed in small but indispensible ways. These would include receptionists, especially at the Colorado River Munici-pal Water District in Big Spring, Texas, and the Panhandle Groundwa-ter Conservation District in White Deer, also in Texas; librarians, especially those in Clayton (New Mexico), Amarillo (Texas), Great Bend (Kansas), and Ogalalla, Alliance, and North Platte (Nebraska); motel staff, especially at the Super 8 motels in Guymon, Oklahoma, and Kadoka, South Dakota, and at the Howard Johnsons in Hobbs, New Mexico; museum and park staff, especially at the American Wind Power Center in Lubbock, Texas, and the Big Well in Greensburg, Kansas; and restaurant staff, especially at My Big Fat Greek Sandwich Shop in North Platte (whose gyros and baklava are world-class) and at Herman's in Big Spring and Roger's in North Platte (which I will rashly declare the two best breakfast spots on the entire High Plains).

Finally, I need to mention the special contributions to the book made by my daughter, Sara Scholz, in San Jose, California, and my "Moroccan daughter," Tiazza Ait Bendaoud (now Tiazza Wilson), in Ashland, Oregon. Sara and her family—husband Howard and children Wolfgang and Emily—hosted us for several days on our way to the High Plains from Oregon, moved a birthday celebration so we could participate, and generally gave us the kind of send-off that got the research on a proper footing. Tiazza took care of the house and the cat while her adopted parents were gallivanting around a distant part of her adopted country, and then put up with all the difficulties and distractions of being part of an author's household over the next two years while I wrote the book and she completed her degree in accounting (with honors) at Southern Oregon University.

Many thanks to all of you, and to the blue Ogallala herself.

Ashland, Oregon
October 12, 2005

BIBLIOGRAPHY

BOOKS

Barlow, Maude, and Tony Clarke. *Blue Gold: The Fight to Stop the Corporate Theft of the World's Water*. New York: New Press, 2002.

Bleed, Ann S., and Charles A. Flowerday, eds. *An Atlas of the Sand Hills*. 3d ed., expanded. Lincoln, NE: University of Nebraska–Lincoln, Institute of Agriculture and Natural Resources Conservation and Survey Division Resource Atlas No. 5b, May 1998.

Brune, Gunnar. *Springs of Texas*. Vol. 1. 2nd ed. College Station, TX: Texas A&M University Press, 2002.

Buchanan, Rex C., and James R. McCauley. *Roadside Kansas: A Traveler's Guide to Its Geology and Landmarks*. Lawrence, KS: University Press of Kansas, 1987.

Clark, John W., Warren Viewssman Jr., and Mark J. Hammer. *Water Supply and Pollution Control*. 3rd ed. New York: Harper & Row, 1977.

Collins, Joseph T., ed. *Natural Kansas*. Lawrence, KS: University Press of Kansas, 1985.

Dumol, Mark. *The Manila Water Concession: A Key Government Official's Diary of the World's Largest Water Privatization*. Washington, DC: World Bank Directions in Development Series, 2000.

Flowerday, Charles A., ed. *Flat Water: A History of Nebraska and Its Water*. Lincoln, NE: University of Nebraska–Lincoln, Institute of Agriculture and Natural Resources Conservation and Survey Division Resource Report No. 12, March 1993.

Frazier, Ian. *Great Plains*. New York: Farrar Straus Giroux, 1989.

Getches, David H. *Water Law in a Nutshell*. St. Paul, MN: West Publishing Company, 1984.

Glennon, Robert. *Water Follies: Groundwater Pumping and the Fate of America's Fresh Waters*. Washington, DC: Island Press, 2002.

Maher, Harmon D., Jr., George F. Engelmann, and Robert D. Shuster. *Roadside Geology of Nebraska*. Missoula, MT: Mountain Press Publishing Company, 2003.

Meinzer, Wyman. *Canyons of the Texas High Plains*. Lubbock, TX: Texas Tech University Press, 2001.

Opie, John. *Ogallala: Water for a Dry Land*. Lincoln, NE: University of Nebraska Press, Our Sustainable Future Series, vol. 1, 1993.

Parkman, Francis. *The Oregon Trail: Sketches of Prairie and Rocky-Mountain Life*. Centenary ed. Boston: Little, Brown and Company, 1922.

Rathjen, Frederick W. *The Texas Panhandle Frontier*. Rev. ed. Lubbock, TX: Texas Tech University Press, 1998.

Reeves, C.C., Jr., and Judy A. Reeves. *The Ogallala Aquifer (of the Southern High Plains)*. Vol. 1. Lubbock, TX: Estacado Books, 1996.

Shiva, Vandana. *Water Wars: Privatization, Pollution and Profit*. Cambridge, MA: South End Press, 2002.

Svobida, Lawrence. *Farming the Dust Bowl: A First-Hand Account from Kansas*. 2nd. ed. Lawrence, KS: University Press of Kansas, 1986.

Wagner, Elsie D. *Cimarron: The Growth of a Town*. [Cimarron, KS?]: self-published, [1976]. Also available online at http://www.cimarronkansas.net/elsie wagner1.htm (31 July 2003).

Ward, Diane Raines. *Water Wars: Drought, Flood, Folly, and the Politics of Thirst*. New York: Riverhead Books, 2002.

Webb, Walter Prescott. *The Great Plains*. 2nd ed. Lincoln, NE: University of Nebraska Press, 1959.

Welsch, Roger. *Sod Walls: The Story of the Nebraska Sod House*. Lincoln, NE: J. & L. Lee Co., 1991.

ARTICLES

(Articles from magazines, journals, and newspapers; newsletter articles and news releases; and individual selections from books.)

American Indian College Fund. "Kellogg Foundation Network Teaches Indians to Raise Buffalo," news release, 1 January 1999. Available online at http://www.charitywire.com/charity4/02218.html (4 August 2003).

Breslau, Karen. "Wildcatting For Water." *Newsweek*, 2 September 2002.

Brooks, Elizabeth, and Jacque Emel. "The Llano Estacado of the American Southern High Plains." In *Regions At Risk: Comparisons Of Threatened Environments*, edited by Jeanne X. Kasperson, Roger E. Kasperson, and B. L. Turner II. Tokyo, Japan: United Nations University Press, 1995. Also available online at http://www.unu.edu/unupress/unupbooks/uu14re/uu14re0n.htm#6.%20the%20llano%20estacado%20of%20the%20american%20southern%20high%20plains (18 February 2004).

Canby, Thomas Y. "Water: Our Most Precious Resource." *National Geographic*, August 1980.

Capps, Riste. "Relief May Be in Sight for South Platte River Water." *News Tribune* (Brush, CO), 30 April 2003.

"Center Pivots Tapping Central Canal System to Irrigate Area Crops." *The Communicator* (Holdrege, NE), March/April 1998.

Cox, Bob. "T. Boone Pickens Takes On Texas Water." *Wichita Eagle* (Wichita, KS), 20 August 2000. Also available online in Mary Hendrickson, *Water, Water*, n.d. http://csf.colorado.edu/archive/2000/food_security/msg00690.html (20 February 2003).

Culver, Gene. "Drilling and Well Construction." In *Geothermal Direct-Use Engineering and Design Guidebook*, edited by J. W. Lund, P. J. Lienau, and B. C. Lunis. Klamath Falls, OR: Geo-Heat Center, Oregon Institute of Technology, 1998.

Deming, David. "Nelson Horatio Darton: Master of Field Geology." In *Introduction to Hydrogeology*, New York: McGraw-Hill, 2002. Available online at http://www.edge.ou.edu/hydrogeology/page10.html (30 May 2003).

"Dwindling Water Supplies Shape Future of Farming in Western Kansas." *U.S. Water News Online*, October 2001. Available at http://www. uswaternews.com/archives/arcsupply/1dwiwat10.html (25 April 2002).

Elder, Robert, Jr. "Range War Over Water Brewing in Panhandle." *American-Statesman* (Austin, TX), 22 February 2003.

Finnegan, William. "Leasing the Rain." *New Yorker*, 8 April 2002.

Glenn, Walter. "Groundwater Conservation District Protects Our Future," news release from the Southeast Texas Groundwater Conservation District, 16 July 2002. Available online at http://www.detcog.org/groundwaterdistrict/districts protectfuture.doc (29 October 2004).

Graham, W. H. "Lea Water Users Hunt for Answers." *Lovington Leader* (Lovington, NM), 16 December 1997.

"Groundwater Depletion Rule Finally Approved." *Panhandle Water News* (White Deer, TX), December 2004.

"Harper County, Oklahoma Canyons Pedestrian Survey." *Oklahoma Archeological Survey Newsletter*, August 2002.

Harriman, Peter. "Bad Water Plagues Tribes." *Argus Leader* (Sioux Falls, SD), 4 November 2001. Available online at http://sdarws.com/SD%20NEWS/2001/November/Nov%204.htm (9 May 2002).

Hobbs, Greg (Gregory J.). "Ground Water Law in Colorado." *Colorado Water*, April 2000.

———. "Colorado Water Law: An Historical Overview." *University of Denver Water Law Review*, Fall 1997. Available online at http://www.law.du.edu/waterlaw/HOBBS_Article.htm (27 January 2003).

Hoelscher, Gail Owens. "The Luckiest Guy in the World." Book review in *Turnarounds and Workouts*, December 15, 2001. Available online at http://www.beardbooks.com/the_ luckiest_guy_in_the_world.html (11 August 2003).

Hofman, Jack L. "The Clovis Hunters: A Pragmatic & Skilled Culture Swept Across North America." *Discovering Archaeology*, January/February 2000. Available online at http://www.panhandlenation.com/prehistory/disc_arc/clovis.htm (4 February 2004).

International Bottled Water Association. "Bottled Water: More Than Just A Story About Sales Growth. Stringent Federal, State and Industry Standards Help Ensure Safety, Quality and Good Taste," news release, 20 May 2003. Also available online at http://www.bottledwater.org/public/2003_Releases/BottledWaterBythe Numbers.htm (5 March 2004).

"In the Great American Desert." *Economist*, 15 December 2001.

"Inventorying Nebraska's Irrigation Acres." *Water Current* (University of Nebraska, Lincoln Water Center), August 2001.

Jehl, Douglas. "Atlanta's Growing Thirst Creates Water War." *New York Times*, 27 May 2002.

———. "Saving Water, U.S. Farmers Are Worried They'll Parch." *New York Times*, 28 August 2002.

King, Philip B. "Memorial to Nelson Horatio Darton." *Proceedings Volume of the Geological Society of America: Annual Report for 1948* (April 1949).

Knickerbocker, Brad. "Privatizing Water: A Glass Half Empty?" *Christian Science Monitor*, 24 October 2002.

LaDuke, Winona. "The Hogs Of Rosebud." *Agribusiness Examiner*, 12 June 2003. Available online at http://www.organicconsumers.org/Toxic/rosebud_hogfarm.cfm (1 October 2003).

"The Land Loan Picture on the Texas High Plains." *The Cross Section* (High Plains Underground Water Conservation District, Lubbock, TX), November 1965.

Lauby, George. "Sandhill Residents Worry About Draining of Aquifer." *North Platte Telegraph* (North Platte, NE), December 21, 2002.

Lavelle, Marianne, Joshua Kurlantzick, and David D'Addio. "The Coming Water Crisis." *U.S. News & World Report*, 12 August 2002.

Leslie, Jacques. "High Noon at the Ogallala Aquifer." *Salon.com*, 1 February 2001. http://www.salon.com/tech/feature/2001/02/01/water_texas/ (25 April 2002).

Lowitt, Richard. "Optima Dam: A Failed Effort to Irrigate the Oklahoma Panhandle." *Agricultural History*, 76, no. 2 (April 2002).

Lutz, Jennifer. "Amarillo's Water Worries: City Ponders Future of Water Availability, Competing with Mesa." *Globe-News* (Amarillo, TX), 9 September 2001.

Mangelsdorf, Martha, and Karen Freiberg. "We're Running Out: A Special Section of Reprinted Articles." *Wichita Eagle and Beacon* (Wichita, KS), February 1979.

Master, Melissa. "Just Another Commodity? If We Treat Water Like One, We Could Come To Grips With The Shortage." *Across the Board*, July/August 2002.

May, Derek. "Water Planners Looking Down." *Avalanche-Journal* (Lubbock, TX), 5 October 2003. Available online at http://www.texaswatermatters.org/pdfs/news_126.pdf (8 December 2004).

McAlavy, Tim, and Pam Dillard. "LEPA Leaps Forward." *Lifescapes*, 3, no. 1 (Spring 2003).

Monroe, Watson H. "Memorial: Nelson Horatio Darton (1865–1948)." *Bulletin of the American Association of Petroleum Geologists*, 33, no. 1 (January 1949).

"More Water, Please: LCWUA Asks 55,000 Acre Feet." *Lovington Leader* (Lovington, NM), 18 August 1999.

Nocera, Joseph. "Return of the Raider." *Fortune*, 27 May 2002.

Oklahoma House of Representatives. "Maximum Yields for Ogallala Aquifer Established by State Water Board," news release, 13 March 2002. Available online at http://www.lsb.state.ok.us/house/news5487.htm (11 March 2003).

Panhandle Groundwater Conservation District. "Panhandle GCD Adopts New Rules, Returns One to Committee," news release, 24 March 2004.

———. "Panhandle GCD Plans to Add 1.25% Limit to Rules: Aquifer Decline Limit to Enforce 50/50 Standard," news release, 12 May 2004.

Patoski, Joe Nick. "Boone Pickens Wants to Sell You His Water." *Texas Monthly*, August 2001. Available online at http://www. texasmonthly.com/mag/issues/2001–08–01/feature6.php (6 June 2002).

Peterson, Jeffrey M., Thomas L. Marsh, and Jeffrey R. Williams. "Conserving the Ogallala Aquifer: Efficiency, Equity, and Moral Motives." *Choices*, February 2003. Available online at http://www.choicesmagazine.org/2003–1/2003-1-04.htm (30 October 2004).

"Pickens Withdraws Request for Freshwater District." The Associated Press, 25 April 2003. Available online at http://www.texaswatermatters.net/pdfs/news_107.pdf (5 May 2003).

Popper, Deborah Epstein, and Frank J. Popper. "The Great Plains: From Dust to Dust. A Daring Proposal for Dealing with an Inevitable Disaster." *Planning*, December 1987. Available online at http://www.planning.org/25anniversary/planning/1987 dec. htm (19 January 2005).

Powers, Tim. "The Slaton Well and the Beginning of the Irrigation Era in Hale County." *Hale County History*, May 1976.

Price, Mike. "Still Gushing: Pickens Sees Fort Worth as Boon to His Water Sales." *Fort Worth Business Press*, 8–14 November 2002.

Ranney, Dave. "'Buffalo Commons' Idea Gets Second Look: Population Decline Rekindles Debate." *Journal-World* (Lawrence, KS), 9 February 2004. Available online at http://www2.ljworld.com/news/2004/feb/09/buffalo_commons_idea/ (19 January 2005).

"Republican River Negotiations with Kansas Come to a Halt." *Nebraska Resources*, Spring 1997. Available online at http://www.dnr.state.ne.us/dnrnews/spring97/ page1.html (14 January 2003).

"Roberts County, Texas, Votes to Accept Findings on Proposed Water District." *Globe-News* (Amarillo, TX), 14 January 2003.

Rothschild, Scott. "Arkansas River Legal Battle Costly for Kansas." *Journal-World* (Lawrence, KS), 12 December 2004. Available online at http://www2.ljworld. com/news/2004/dec/12/arkansas_river_legal/ (23 December 2004).

Russell, Daniel. "Move Made to Protect Lea Water Rights." *News Sun* (Hobbs, NM), 19 August 1999.

Salcetti, Marianne. "Liquid Assets." *Garden City Telegram* (Garden City, KS), 21 October 2000. Available online at http://www.gctelegram.com/news/2000/October /21/liquidassets.html (28 October 2004).

"Sale of Water to Colorado Targeted." *Wise Water Words* (Nebraska Section, American Water Works Association), Spring 2003.

Shela, Jason G. "Boone Pickens, a Man For All Seasons. Part I: The Early Years." *Opportunity* (Stern School of Business, New York University), no. 14 (1998–

1999). Available online at http://pages.stern.nyu.edu/~opportun/issues/1998-99/issue14. htm (8 August 2003).

Shirley, Kathy. "Pickens Jumps Into Water Market." *Explorer* (American Association of Petroleum Geologists), July 2001. Available online at http://www.aapg.org/explorer/2001/07jul/pickens_water.cfm (20 February 2003).

Simpson, Hal D. "Administration of Ground Water in Colorado." *Colorado Water*, April 2000.

Slaton, Mozelle Marlin. "John Henry Slaton." *Hale County History* (Hale County, TX), May 1976.

Storm, Rick. "Water Wars, Part One: An Old Story." *Globe-News* (Amarillo, TX), 30 June 2002.

———. "Water Wars, Part Two: Another Landowners' Group Eyes Roberts County Water." *Globe-News* (Amarillo, TX), 1 July 2002.

"Subunit Management Will Soon Become Public Policy." *Water District Newsletter* (Southwest Kansas Groundwater Management District, Garden City, KS), March 2004.

"Texas Groundwater: Yours? Mine? Ours?" *Panhandle Water News* (White Deer, TX), July 2003.

"Texas Tech Study Assesses How Depletion of the Ogallala Aquifer May Impact Farmers, Counties in the High Plains." *NewWaves* (Texas Tech Water Resources Institute, Lubbock, TX), June 2001.

Thorpe, Helen. "Waterworld." *Texas Monthly*, September 1995.

Wallach, Brett. "The Physical Environment of the Great Plains." In *Encyclopedia of the Great Plains*, edited by David Wishart. Lincoln, NE: University of Nebraska Press, 2004.

Wark, Andrew. "Spearhead Discovery Puts Horse On Prehistoric Menu." News release from the University of Calgary, May 2, 2001. Available online at http://www.fp.ucalgary.ca/unicomm/news/Stmary/spearhead.htm (5 February 2004).

"Water Resources and Planning: Sicangu Mni Wiconi Project." In MSE Technology Applications Inc., *1998 MSE Annual Report*. Available online at http://www.mse-ta.com/news/reports/1998annual.pdf (1 October 2003).

Willis, John A. K. "Paleo-Indians." In *Encyclopedia of North American Indians*, edited by Frederick E Hoxie. Boston: Houghton Mifflin, 1996. Available online at http://college.hmco.com/history/readerscomp/naind/html/na_027500_paleo indians.htm (4 February 2004).

Yardley, Jim. "For Texas Now, Water and Not Oil Is Liquid Gold." *New York Times*,

April 16, 2001. Available online at http://www.rioweb.org/Archive/nyt_water oi1041601.html (20 February 2003).

Yung, Katherine. "Corporate Raider Deals Again, but This Time the Commodity Is Water." *Morning News* (Dallas, TX), 3 November 2002.

Zwingle, Erla. "Wellspring of the High Plains." *National Geographic*, March 1993.

TECHNICAL MATERIALS

(Planning documents, legal documents, and research papers.)

Agency for Toxic Substances and Disease Registry. *Public Health Assessment, Ace Services Incorporated, Colby, Thomas County, Kansas.* Atlanta, GA: U.S. Department of Health and Human Services, Public Health Service Cerclis No. Ksd046746731, June 14, 1996. Available online at http://www.atsdr.cdc.gov/HAC/PHA/ace_serv/ace_toc.htm (23 September 2004).

Article 9: Underground Water. Wyoming Statutes, 41-3-9.

Becker, Mark F., Breton W. Bruce, Larry M. Pope, and William J. Andrews. *Ground-Water Quality in the Central High Plains Aquifer, Colorado, Kansas, New Mexico, Oklahoma, and Texas, 1999.* Oklahoma City, OK: U.S. Geological Survey National Water-Quality Assessment Program Water-Resources Investigations Report 02-4112, 2002.

Brown, J. E. "Buster." *Senate Bill 2.* Available online at http://www.texaswater.org/water/law/sb2abs.htm (27 January 2003).

CH2M Hill, Inc. *Technical Report: Central's Irrigation Division Water Conservation and Management Program.* Prepared for Central Nebraska Public Power and Irrigation District. Denver: CH2M Hill, Inc., May 1, 1991.

Clyma, Wayne, and F. B. Lotspeich. *Occurrence and Use of Water Resources in the High Plains of Texas and New Mexico. For Presentation at the 1963 Winter Meeting, American Society of Agricultural Engineers, Chicago, Illinois December 10–13.* St. Joseph, MI: American Society of Agricultural Engineers Paper No. 63-741, 1963.

Darton, Nelson Horatio. *Preliminary Report on the Geology and Water Resources of Nebraska West of the One Hundred and Third Meridian.* Washington, DC: U.S. Geological Survey Professional Paper No. 17, 1903.

Dvorak, Bruce I., DeLynn Hay, and Bill Welton. *Technical Assistance and Education for the Native American Nations In Kansas, Nebraska and South Dakota.* Champaign, IL: Midwest Technology Assistance Center for Small Public Water Systems, 25 January 2000. Available online at http://mtac.sws.uiuc.edu/mtacdocs/NatAm FinRpt/NatAmFinRpt00.pdf (21 May 2003).

Gilson, Preston, Joseph A. Aistrup, John Heinrichs, and Brett Zollinger. *The Value of Ogallala Aquifer Water in Southwest Kansas. Prepared for Southwest Kansas Groundwater Management District.* Hays, KS: The Docking Institute of Public Affairs, Fort Hays State University, 2001.

Gisser, Micha. *Economic Aspects of Water in New Mexico.* Tijeras, NM: The Rio Grande Foundation, n.d. Available online at http://www.riograndefoundation.org/pa pers/economic_aspects_of_water.htm (2 June 2004).

Gutentag, Edwin D., Frederick J. Heimes, Noel C. Krothe, Richard R. Luckey, and John B. Weeks. *Geohydrology of the High Plains Aquifer in Parts of Colorado, Kansas, Nebraska, New Mexico, Oklahoma, South Dakota, Texas, and Wyoming.* Washington, DC: U.S. Geological Survey Professional Paper 1400-B, High Plains RASA Project, 1984.

Hofman, Jack L., and India S. Hesse. "The Occurrence of Clovis Points in Kansas." *Current Research in the Pleistocene* 13:23–25 (1996). Available online at http:// www.peak.org/csfa/crp13-11.html (4 February 2004).

Hopkins, Janie. *Water-Quality Evaluation of the Ogallala Aquifer, Texas.* Austin, TX: Texas Water Development Board Report 342, August 1993.

Howell, Terry A., Judy A. Tolk, Steve R. Evett, and R. Louis Baumardt. *Cotton and Sorghum Rotation Under Deficit Furrow Irrigation.* St. Joseph, MI: American Society of Agricultural Engineers Meeting Paper No. 032134, 2003.

Kansas Water Authority. *The Kansas Water Plan: Fiscal Year 2003.* [Topeka, KS?]: Kansas Water Office, July 2001.

Kansas Water Resources Board. *Ground Water Problems in Southwestern Kansas. Interim Report to the Water Resources Committees of the Kansas Legislature.* Topeka, KS: Kansas Water Resources Board, January 1959.

Layher, Bill. *Recovery Plan for the Scott Riffle Beetle, Optioservus phaeus Gilbert, in Kansas.* Topeka, KS: Kansas Department of Wildlife and Parks, April 2002.

Luckey, Richard R., Edwin D. Gutentag, Frederick J. Heimes, and John B. Weeks. *Effects of Future Ground-Water Pumpage on the High Plains Aquifer in Parts of Colorado, Kansas, Nebraska, New Mexico, Oklahoma, South Dakota, Texas, and Wyoming.* Washington, DC: U.S. Geological Survey Professional Paper 1400-E, High Plains RASA Project, 1988.

———. *Digital Simulation of Ground-Water Flow in the High Plains Aquifer in Parts of Colorado, Kansas, Nebraska, New Mexico, Oklahoma, South Dakota, Texas, and Wyoming.* Washington, DC: U.S. Geological Survey Professional Paper 1400-D, High Plains RASA Project, 1986.

McGuire, V. L., M. R. Johnson, R. L. Schieffer, J. S. Stanton, S. K. Sebree, and I. M. Verstraeten. *Water in Storage and Approaches to Ground-Water Management, High Plains Aquifer, 2000.* Reston, VA: U.S. Geological Survey Circular 1243, 2003.

Nebraska Natural Resources Commission. *Policy Issue Study on Integrated Management of Surface Water and Groundwater.* [Lincoln, NE?]: Nebraska Natural Resources Commission, April 1986.

New Mexico Energy, Minerals and Natural Resources Department, State Parks Division. *Oasis State Park Management And Development Plan.* [Portales, NM?]: New Mexico Energy, Minerals and Natural Resources Department, n.d. Available online at http://www.emnrd.state.nm.us/nmparks/PAGES/mgmtplans/oasis mgmtplan.pdf (12 February 2003).

New Mexico Office of the State Engineer and Interstate Stream Commission. *Region 16—Lea County Regional Water Plan,* 24 August 2003. Available online at http://www.ose.state.nm.us/water-info/NMWaterPlanning/regions/leacounty/leacnty-menu.html (13 October 2004).

———. *2001–2002 Annual Report.* Santa Fe, NM: Office of the State Engineer and Interstate Stream Commission, 1 April 2003. Available online at http://www.ose .state.nm.us/publications/01-02-annual-report/toc.html (13 October 2004).

———. *1998–1999 Annual Report.* Santa Fe, NM: Office of the State Engineer and Interstate Stream Commission, 4 April 2000. Available online at http://www.ose.state.nm.us/publications/98–99-annual-report/index.html (13 October 2004).

———. "Ground Water in Curry and Roosevelt Counties, New Mexico." In *Tenth Biennial Report.* Albuquerque, NM: State Engineer, 1930–32.

Northwest Kansas Groundwater Management District No. 4. *Revised Management Program.* Colby, KS: Northwest Kansas Groundwater Management District No. 4, 1 May 1994.

Ogallala Aquifer Management Advisory Committee. *Discussion and Recommendations for Long-Term Management of the Ogallala Aquifer in Kansas.* Topeka, KS: Kansas Water Office, 16 October 2001.

Oklahoma Water Resources Board. *Chapter 10: Special Purpose Districts.* Oklahoma Water Resources Board, 1998. Available online at http://www.owrb. state.ok.us/ util/rules/pdf_rul/Chap10.pdf (1 November 2004).

———. *Chapter 30: Taking and Use of Groundwater: Final Proposed Amendments.* Oklahoma City, OK: Oklahoma Water Resources Board, 10 February 2004. Available online at http://www.owrb.state.ok.us/util/rules/pdf_rul/draft_ch30.pdf (1 November 2004).

———. *Update of the Oklahoma Comprehensive Water Plan.* Oklahoma Water Resources Board, 1995. Available online at http://www.owrb.state.ok.us/supply/ocwp/ocwp1995.php (11 March 2003).

Panhandle Groundwater Conservation District. *Draft Rules of Panhandle Groundwater Conservation District, in Texas, as Amended.* White Deer, TX: Panhandle Groundwater Conservation District, 20 December 2003.

———. *Panhandle Groundwater Conservation District Annual Report 2001–2002.* White Deer, TX: Panhandle Groundwater Conservation District, 30 September 2002.

———. *Panhandle Groundwater Conservation District Management Plan.* White Deer, TX: Panhandle Groundwater Conservation District, 31 December 2002.

———. *Panhandle Groundwater Conservation District Management Plan.* White Deer, TX: Panhandle Groundwater Conservation District, 1998.

———. *Rules of Panhandle Ground Water Conservation District No. 3, in Texas, as Amended.* White Deer, TX: Panhandle Groundwater Conservation District, 18 March 1998.

Panhandle Water Planning Group. "Executive Summary." In *Regional Water Plan: Panhandle Water Planning Area.* Austin, TX: Texas Water Development Board, [2000?]. Available online at http://www.twdb.state.tx.us/rwp/a/PDFs/A_Executive%20Summary.pdf (27 January 2003).

Peterson, Steven M., and Clint P. Carney. *Estimated Groundwater Discharge to Streams from the High Plains Aquifer in the Eastern Model Unit of the COHYST Study Area for the Period Prior to Major Groundwater Irrigation.* [Lincoln, NE?]: Cooperative Hydrology Study, 14 March 2002.

Popper, Deborah E., and Frank J. Popper. *The Buffalo Commons as Regional Metaphor and Geographic Method.* Great Plains Restoration Council, Fort Worth, TX, and Denver, CO, n.d. Available online at http://www.gprc.org/buffalo_commons_popper.html (19 January 2005).

Rainwater, Ken, and David B. Thompson. "Playa Lake Influence on Ground-Water Mounding in Lubbock, Texas." In *Proceedings of the Playa Basin Symposium, 1994,* edited by Lloyd V. Urban and A. Wayne Wyatt. Lubbock, TX: Texas Tech University, 1994.

Rattlesnake Creek/Quivira Partnership. *Rattlesnake Creek Management Program Proposal.* Stafford, KS: Rattlesnake Creek/Quivira Partnership, 29 June 2000.

Republican River Compact. Kansas Statutes, 82a-518.

Robertson, Paul M. "Identity, Class, and New Deal Politics on Pine Ridge." In *The*

Power of the Land, unpub. Ph.D. dissertation, 1995. Available online at http://
www.fireonprairie.org/Colonization%20Pine%20Ridge.html (4 August 2003).

Sawin, Robert, Rex Buchanan, and Wayne Lebsack. *Kansas Springs Inventory: Water
Quality, Flow Rate, and Temperature Data.* [Topeka, KS?]: Kansas Geological Survey Open File Report 2002-46, November 2002. Available online at http://www
.kgs.ku.edu/Hydro/Publications/OFR02_46/index.html (20 September 2004).

Soenksen, Philip J., Lisa D. Miller, Jennifer B. Sharpe, and Jason R. Watton. "Appendix
A: Descriptions of Selected Drainage-Basin Characteristics Quantified Using
Basinsoft, Arc-Info, and Related GIS Programs." In *Peak-Flow Frequency Rela-
tions and Evaluation of the Peak-Flow Gaging Network in Nebraska.* [Lincoln,
NE?]: U.S. Geological Survey Water Science Center Water-Resources Investiga-
tions Report 99-4032, 1999.

Southwest Kansas Groundwater Management District. *2004 Revised Management
Program, Draft Version.* Garden City, KS: Southwest Kansas Groundwater Management District, January 14, 2004.

Swanson, R. B., E. J. Blajszczak, S. C. Roberts, K. R. Watson, and J. P. Mason. *Water
Resources Data, Wyoming, Water Year 2002.* Vol. 2, *Groundwater.* Cheyenne, WY:
U.S. Geological Survey Water-Data Report WY-02-2, 2003.

Sweeten, John M., and Wayne R. Jordan. *Irrigation Water Management for the Texas
High Plains: A Research Summary.* College Station, TX: Texas A&M University.
Texas Water Resources Institute Technical Report No. 139, August, 1987.

Templer, Otis W. "Municipal Conjunctive Water Use on the Texas High Plains." *Social
Science Journal,* 38 (2001).

Templer, Otis W., and Lloyd V. Urban. "Conjunctive Use of Water on the Texas High
Plains." *Journal of Contemporary Water Research and Education,* 106 (Winter
1996). Available online at http://www.ucowr. siu.edu/updates/pdf/V106_A13.pdf
(8 December 2004).

Texas Water Development Board. *Water for Texas—2002* [state water plan]. Austin,
TX: Texas Water Development Board, 2002.

U.S. Geological Survey. *Water Resources Investigations in Kansas, 1965. Conducted by
the United States Geological Survey Water Resources Division in Cooperation with
State, Municipal, and Federal Agencies.* Topeka, KS: U.S. Geological Survey, 1965.

Water Conservation Districts. Texas Water Code, Title 2, Chapter 36. Available online
at   http://www.capitol.state.tx.us/statutes/docs/WA/content/htm/wa.002.00.000
036.00. htm (29 October 2004).

Weeks, John B., Edwin D. Gutentag, Frederick J. Heimes, and Richard R. Luckey. *Sum-

mary of the High Plains Regional Aquifer-System Analysis In Parts of Colorado, Kansas, Nebraska, New Mexico, Oklahoma, South Dakota, Texas, and Wyoming. Washington, DC: U.S. Geological Survey Professional Paper 1400-A, High Plains RASA Project, 1988.

Western Kansas Groundwater Management District No. 1. *Management Program.* Scott City, KS: Western Kansas Groundwater Management District No. 1, n.d. Available online at http://www.gmd1.org/ManagementProgram.htm (20 September 2004).

Whetstone, George A., ed. *Proceedings of the Ogallala Aquifer Symposium II.* Lubbock, TX: Texas Tech University Water Resources Center, 1984.

Wiersma, Ursula M., Victor R. Hasfurther, and Greg L. Kerr. "Tectonic Structure." In *Structural Obstruction of Recharge to the Paleozoic Aquifer in the Denver-Julesburg Basin along the Laramie Range, Wyoming.* Laramie, WY: University of Wyoming, Wyoming Water Research Center Research Project Technical Completion Report WWRC-88-04, May 1989. Available online at http://library.wrds.uwyo.edu/wrp/88-04/ch-04.html (1 May 2003).

Wood, Warren W. "Role of Ground Water in Geomorphology, Geology, and Paleoclimate of the Southern High Plains, USA." *Ground-Water*, 40, no. 4 (July–August 2002).

Wood, Warren W., Stephen Stokes, and Julie Rich. "Implications of Water Supply for Indigenous Americans during Holocene Aridity Phases on the Southern High Plains, USA." *Quarternary Research*, 58 (2002).

Woodward, Dennis. "The High Plains (Ogallala) Aquifer: Managing the Resource in the Southern High Plains, New Mexico." In *Proceedings of the 42nd Annual New Mexico Water Conference: Water Issues of Eastern New Mexico.* Las Cruces, NM: New Mexico State University, New Mexico Water Resources Research Institute Report No. 304, 1997. Available online at http://wrri.nmsu.edu/publish/watcon/proc42/woodward.html (17 January 2003).

MISCELLANEOUS PRINTED MATERIALS
(Booklets, pamphlets, brochures, fact sheets, and maps.)

Aldrich, Janette Marie, Wayne R. Ostlie, and Thomas M. Faust. *The Status of Biodiversity in the Great Plains: Great Plains Landscapes of Biological Significance.* Arlington, VA: The Nature Conservancy, 24 April 1997.

American Pima Cotton: Acres Planted—2002. Austin, TX: Texas Field Office, National

Agricultural Statistics Service, 2003. Available online at http://www.nass.usda
.gov/tx/bu02_085.pdf (29 July 2004).

Amosson, Steve, Leon New, Lal Almas, Fran Bretz, and Thomas Marek. *Economics of
Irrigation Systems*. Texas Cooperative Extension Bulletin B-6113, December
2001.

Ashworth, John B. *Water-Level Changes in the High Plains Aquifer of Texas*. [Austin,
TX?]: Texas Water Development Board Hydrologic Atlas No. 1, 1991.

Atmospheric Water. Bismarck, ND: North Dakota Water Education Foundation, October 2000.

Buettner, Jeff. *A Journey Through the Central District: A Comprehensive Summary of
the Origin, Development, Facilities and Contributions of the Central Nebraska
Public Power and Irrigation District*. 2nd ed. Holdrege, NE: Central Nebraska
Public Power and Irrigation District, November 2001.

Carlson, Marvin P. *Geology, Geologic Time and Nebraska*. Lincoln, NE: University of
Nebraska–Lincoln, Institute of Agriculture and Natural Resources Conservation
and Survey Division Educational Circular No. 10, August 1993.

*Central Nebraska Public Power and Irrigation District: Irrigation, Hydropower, Recre-
ation, Wildlife Habitat*. Holdrege, NE: Central Nebraska Public Power and Irrigation District, n.d.

Cimarron Chamber of Commerce. *Cimarron Kansas History*, 10 April 2002. Available
online at http://skyways.lib.ks.us/towns/Cimarron/history.html (31 July 2003).

Cruz, R. R. *Groundwater Levels, Clovis Area, New Mexico, 1982–1987*. Albuquerque,
NM: New Mexico State Engineer's Office Map No. GWL-CA-82/87, 1989.

Eastern New Mexico University. *Blackwater Draw Museum*. Portales, NM: Eastern
New Mexico University, n.d.

Farrar, Jon. *A Wildflower Year*. Lincoln, NE: Nebraska Game and Parks Commission
Nebraskaland magazine, 68, no. 1, 1990.

Fipps, Guy. *Managing Texas' Groundwater Resources through Groundwater Conservation Districts*. College Station, TX: Texas A&M University System, Texas Agricultural Extension Service Brochure B-1612, November 1998.

Flowerday, Charles A., ed. *Groundwater Atlas of Nebraska*. 2nd rev. ed. Lincoln, NE:
University of Nebraska–Lincoln, Institute of Agriculture and Natural Resources
Conservation and Survey Division Resource Atlas No. 4a, 1998.

Garrabrant, Lynn A. *Water Use in New Mexico, 1990*. Albuquerque, NM: U.S. Geological Survey, 1994.

Gilkerson, Joni G., and John R. Bozell. *Historic Places: The National Register for*

Nebraska. Lincoln, NE: Nebraska Game and Parks Commission *Nebraskaland* magazine, 67, no. 1, 1989.

Hansen, Hank. *Methods of Dealing with Groundwater Supply Problems*. Garden City, KS: Southwest Kansas Groundwater Management District No. 3, 8 October 2002.

High Plains Associates. *High Plains Ogallala Aquifer Regional Study*. Austin, TX: High Plains Associates, n.d.

Jacobs, James J., Patrick T. Tyrrell, and Donald J. Brosz. *Wyoming Water Law: A Summary*. Laramie, WY: University of Wyoming Agricultural Experiment Station Bulletin B-849R, May 2003.

Jones, Ray. *A Summary of the 2002 Weather Modification Program for Rain Enhancement and a Review of Area Rainfall*. Big Spring, TX: Colorado River Municipal Water District, [2003?].

Kansas Department of Wildlife and Parks. *Cheyenne Bottoms Wildlife Area: General Information*. Topeka, KS: Kansas Department of Wildlife and Parks, March 1999.

———. *Kiowa State Fishing Lake*. Topeka, KS: Kansas Department of Wildlife and Parks, June 2000.

———. *Lake Meade State Park*. Topeka, KS: Kansas Department of Wildlife and Parks, June 2000.

———. *Lake Scott State Park and Wildlife Area: El Cuartelejo Indian Pueblo*. Topeka, KS: Kansas Department of Wildlife and Parks, March 2001.

Kansas Groundwater Management Districts Association. *Groundwater Management in Kansas*. [Halstead, KS?]: Kansas Groundwater Management Districts Association, 1980.

Katz, Lienke. *The History of Blackwater Draw*. Portales, NM: Eastern New Mexico University Printing Services, 1997.

Luckey, Richard R., Noel I. Osborn, Mark F. Becker, and William J. Andrews. *Water Flow in the High Plains Aquifer in Northwestern Oklahoma*. [Oklahoma City, OK?]: U.S. Geological Survey Fact Sheet 081-00, June 2000.

Martin, Gene, and Mary Martin. *Trail Dust: A Quick Picture History of the Santa Fe Trail*. 8th ed. Manitou Springs, CO: Martin Associates, 1972.

McGuire, V. L. (Virginia L.), C. P. Stanton, and B. C. Fischer. *Water-Level Changes, 1980 to 1997, and Saturated Thickness, 1996–97, in the High Plains Aquifer*. Reston, VA: U.S. Geological Survey Fact Sheet FS-124-99, [1999?].

McGuire, V. L. (Virginia L.), C. P. Stanton, B. C. Fischer, and Jennifer B. Sharpe. *Water-Level Changes in the High Plains Aquifer, 1980 to 1995*. Reston, VA: U.S. Geological Survey Fact Sheet FS-068-97, July 1997.

Mesa Water, Inc. *Mesa* [PowerPoint presentation]. Dallas, TX: Mesa Water, Inc., n.d.

Misgna, Girmay, Robert Buddemeier, Jeff Schloss, and Keith Lebbin. *Percent Change in Saturated Thickness of the High Plains Aquifer, West Central Kansas 1950 to Average 1997–1999.* [Topeka, KS?]: Kansas Geological Survey Open File Report 2000-15B, [2000?]. Available online at http://www.gmd1.org/images/plate_b_250k. gif (20 September 2004).

———. *Saturated Thickness of Unconsolidated Aquifer, West Central Kansas Average, 1997–1999.* [Topeka, KS?]: Kansas Geological Survey Open File Report 2000-15A, [2000?]. Available online at http://www.gmd1.org/images/plate_a_250k.gif (20 September 2004).

Museum of the Llano Estacado. *Museum of the Llano Estacado.* Plainview, TX: [Wayland Baptist University?], n.d.

National Agricultural Statistics Service. *2002 Census of Agriculture: County Profile: Dallam, Texas.* Washington, DC: National Agricultural Statistics Service, [2003?].

———. *2002 Census of Agriculture: County Profile: Moore, Texas.* Washington, DC: National Agricultural Statistics Service, [2003?].

———. *2002 Census of Agriculture: State Profile: Texas.* Washington, DC: National Agricultural Statistics Service, [2003?].

National Park Service. *Scotts Bluff Official Map and Guide.* Washington, DC: U.S. Government Printing Office, 2002.

Nebraska Department of Natural Resources. *Summary of Republican River Compact Litigation Settlement.* Lincoln, NE: Nebraska Department of Natural Resources, June 2003.

New Mexico Office of the State Engineer and the Interstate Stream Commission. *Can You Tell Me About Domestic Wells in New Mexico?* Santa Fe, NM: Office of the State Engineer and the Interstate Stream Commission, n.d.

———. *Fact Sheet: New Mexico Water Law.* Santa Fe, NM: Office of the State Engineer and the Interstate Stream Commission, n.d.

———. *Regional Water Planning in New Mexico.* Santa Fe, NM: Office of the State Engineer, n.d. Available online at http://www.ose.state.nm.us/doing-business/water-plan/rwpnm-pamphlet.html (1 November 2004).

New Mexico State Parks Division. *Oasis State Park.* Santa Fe, NM: New Mexico State Parks Division, n.d.

New, Leon. *Corn Production Per Inch of Water.* Amarillo, TX: Texas Cooperative Extension, n.d.

———. *Irrigation Pumping Plant Tests.* Amarillo, TX: Texas Cooperative Extension, n.d.

———. *Seasonal Irrigation Capacity*. Amarillo, TX: Texas Cooperative Extension, n.d.

———. *Soybean Production Per Inch of Water*. Amarillo, TX: Texas Cooperative Extension, n.d.

New, Leon, and Guy Fipps. *Center Pivot Irrigation*. Texas Cooperative Extension Bulletin B-6096, April 2000.

Nielsen, D. C. *Crop Rotation, Soil Water Content and Wheat Yields*. Akron, CO: Central Great Plains Research Station Conservation Tillage Fact Sheet #1-02, [2002?].

O'Brien, Daniel M., Troy J. Dumler, and Danny H. Rogers. *Irrigation Capital Requirements and Energy Costs*. [Manhattan, KS?]: Kansas State University, October 2002.

O'Brien, Erin. *Exploring the Sustainability of the Ogallala Aquifer* [PowerPoint presentation]. [Manhattan, KS?]: Kansas State University, 2001.

Panhandle Groundwater Conservation District. *1999 Ogallala Water Table Elevation*. White Deer, TX: Panhandle Groundwater Conservation District, 1999.

———. *Panhandle Groundwater Conservation District*. White Deer, TX: Panhandle Groundwater Conservation District, n.d.

———. *Panhandle Groundwater Conservation District Aquifer Facts*. White Deer, TX: Panhandle Groundwater Conservation District, n.d.

———. *Panhandle Regional Aquifers*. White Deer, TX: Panhandle Regional Groundwater Conservation District, n.d.

Peanuts: Acres Planted—2002. Austin, TX: Texas Field Office, National Agricultural Statistics Service, 2003. Available online at http://www.nass.usda.gov/tx/bu02_100.pdf (29 July 2004).

Phelps County Historical Society. *Nebraska Prairie Museum of the Phelps County Historical Society*. Holdrege, NE: Phelps County Historical Society, n.d.

Porter, Dana, Russell Persyn, and Juan Enciso. *Groundwater Conservation Districts: Success Stories*. [College Place, TX?]: Texas A&M University System, Texas Agricultural Extension Service Brochure B-6087, August 1999.

Prairie Museum of Art and History, Colby, Kansas. Colby, KS: Prairie Museum of Art and History, n.d.

Schlageck, John, ed. *Water Issue(s)*. Special issue of Kansas Farm Bureau *Kansas Living* magazine, Summer 2004.

Schmidt, Shiela Sutton. *Pawnee Rock: A Brief History of the Rock*. Pawnee Rock, KS: May 1986.

Schwarz, Tom. *A Farmer's Guide to Water Rights: Your Rights, Responsibilities and a Guide to Preparing for the Adjudication Process*. Kearney, NE: Nebraska Water Users, Inc., 1993.

Steelquist, Robert U. *Field Guide to the North American Bison: A Natural History and Viewing Guide to the Great Plains Buffalo*. Seattle: Sasquatch Books, 1998.

Texas Alliance of Groundwater Districts Operations Manual Committee. *Groundwater Conservation District Operations Manual*. [Austin, TX?]: Texas Water Development Board, Texas Natural Resource Conservation Commission and Texas Alliance of Groundwater Districts, July 1999. Available online at http://www.texas groundwater.org/rules&regs/Operation%20Manual/WordPerfect%20Version/A _Cover%20and%20Text/03%20OP%20Manual%20Text.htm (29 October 2004).

Trimble, Donald E. *The Geologic Story of the Great Plains: A Nontechnical Description of the Origin and Evolution of the Landscape of the Great Plains*. Washington, DC: U.S. Geological Survey Bulletin 1493, 1980.

U.S. Environmental Protection Agency. *Ace Services Record of Decision Signed*. Kansas City, KS: EPA Region 7, August 1999.

U.S. Fish and Wildlife Service. *Quivira National Wildlife Refuge*. Stafford, KS: U.S. Fish and Wildlife Service, September 2001.

Waite, Herbert A. *Geology and Ground-Water Resources of Ford County, Kansas*. Topeka, KS: Kansas Geological Survey Bulletin 43, December 1942. Available online at http://www.kgs.ku.edu/General/Geology/Ford/index.html (7 September 2004).

Water and the Future of Rural Texas. Austin, TX: Texas Center for Policy Studies, 2001.

Watts, Darrell G., DeLynn R. Hay, and David A. Eigenberg, eds. *Managing Irrigation and Nitrogen to Protect Water Quality*. Lincoln, NE: University of Nebraska Cooperative Extension Circular EC98-786-S, [1998?].

Western Kansas Groundwater Management District No. 1. *Programs*. Scott City, KS: Western Kansas Groundwater Management District No. 1, n.d. Available online at http://www.gmd1.org/Programs.htm (20 September 2004).

Wofford, Vera Dean, ed. *Hale County Facts and Folklore*. Lubbock, TX: Pica Publishing Company, 1978.

Wood, Warren W. *Ground-Water Recharge in the Southern High Plains of Texas and New Mexico*. Reston, VA: U.S. Geological Survey Fact Sheet FS-127-99, August 2000.

WEB PAGES

Agricultural Marketing Service. *Gaines County, TX: Information based on Congressional Districts of the 106th U.S. Congress, 1997*. Agricultural Marketing Service,

USDA, n.d. http://www.ams.usda.gov/statesummaries/TX/County/County_pdf/ Gaines.pdf (13 August 2004).

―――. *Hale County, TX: Information based on Congressional Districts of the 106th U.S. Congress, 1997*. Agricultural Marketing Service, USDA, n.d. http://www .ams.usda.gov/statesummaries/TX/County/County_pdf/Hale.pdf (30 July 2004).

―――. *Lubbock County, TX: Information based on Congressional Districts of the 106th U.S. Congress, 1997*. Agricultural Marketing Service, USDA, n.d. http://www.ams .usda.gov/statesummaries/TX/County/County_pdf/Lubbock.pdf (30 July 2004).

Blackwater Draw Site. Eastern New Mexico State University, 20 July 2002. http:// www.enmu.edu/academics/excellence/museums/blackwater-draw/site/index. shtml (10 March 2003).

Buddy Holly Center. *Buddy Holly Exhibition: Biography*. Buddy Holly Center, 1999. http://www.buddyhollycenter.org/htm/Buddy%20Holly%20Exhibit/bhe_b.htm (28 July 2004).

Calvert, J.B. *The Llano Estacado*. 22 May 2001. http://www.du.edu/~jcalvert/geol/ llano.htm (20 May 2002).

Colorado Ground Water Commission. *Designated Basins and Ground Water Management Districts*. Colorado Ground Water Commission, n.d. http://water.state.co .us/cgwc/DB-GWMgmtDist.htm (1 November 2004).

European Bentonite Producers Association. *What Is Bentonite?* European Bentonite Producers Association, n.d. http://www.ima-eu.org/en/whabentontext.htm (1 August 2003).

Flynn, Keli. "T. Boone Pickens." Famous Texans, n.d. http://www.famoustexans.com/ boonepickens.htm (11 August 2003).

Industrial Minerals Association—North America. *What is Bentonite?* Industrial Minerals Association—North America, n.d. http://www.ima-na.org/about_industrial _minerals/bentonite.asp (1 August 2003)

Jasinski, Laurie E. "Underground Water Conservation Districts." Handbook of Texas Online, 6 June 2001. http://www.tsha.utexas.edu/handbook/online/articles/ UU/mwund.html (29 October 2004).

Kansas Department of Agriculture, Division of Water Resources. *Republican River Compact Background and Update*. Kansas DWR, 18 December 2002. URL not recorded (14 January 2003). No longer online: version of 7 September 2004 at http://www.ksda.gov/Default.aspx?tabid=334 (29 September 2005).

Kansas Fact and Fancy: Trivia Questions About Kansas History. The Kansas Collection, n.d. http://www.kancoll.org/articles/kstrivia/answers2.htm (31 July 2003).

Koontz, John E. *Etymology*, 2 June 2001. http://spot.colorado.edu/~koontz/faq/ety
mology.htm (1 September 2004).

Oklahoma Geological Survey. *Tertiary Rocks of Oklahoma*. Oklahoma Geological Survey, n.d. http://www.ogs.ou.edu/earthscience/intgeol/10tert.htm (11 March 2003).

Oklahoma Water Resources Board. *Frequently Asked Questions*. Oklahoma Water Resources Board, 20 April 2004. http://www.owrb.state.ok.us/util/faq.php (1 November 2004).

———. *Groundwater Permitting*. Oklahoma Water Resources Board, 15 June 2004. http://www.owrb.state.ok.us/supply/watuse/gwwateruse.php (1 November 2004).

Public Employees for Environmental Responsibility. *Pantex: Pollution in the Panhandle*. Toxic Texas, [2000?]. http://www.txpeer.org/toxictour/pantex.html (12 August 2003).

Record, Ian, and Anne Pearse Hocker. *It's All About the Land: The Issue Behind Wounded Knee*. Crazyoglala's Tipi, 17 January 2002. http://www.geocities.com/crazyoglala/WK73_AboutLand_Record.html (4 August 2003).

Romanek, Andrew. *Interesting Statistics of the Ogallala Aquifer and High Plains Region*, Fall 1997. http://www.ce.utexas.edu/prof/maidment/grad/romanek/wtrproject/stats.htm (29 April 2002).

South Dakota Department of Environment and Natural Resources. *Drilling a Well*. South Dakota DENR, 17 March 2004. http://www.state.sd.us/denr/des/water rights/wr_well.htm (1 November 2004).

———. *Frequently Asked Questions about Water Rights*. South Dakota DENR, 17 March 2004. http://www.state.sd.us/denr/des/waterrights/faqwr.htm (1 November 2004).

———. *Summary of SD Water Laws & Rules*. South Dakota DENR, 17 March 2004. http://www.state.sd.us/denr/des/waterrights/summary.htm (1 November 2004).

———. *Using Water in South Dakota*. South Dakota DENR, 17 March 2004. http://www.state.sd.us/denr/des/waterrights/wr_permit.htm (1 November 2004).

South Dakota Public Broadcasting. *Mni Wiconi Water Pipeline*. South Dakota Public Broadcasting, 11 August 2003. http://www. sdpb.org/www/RADIO/news/mni wiconi/index.htm (1 October 2003).

U.S. Environmental Protection Agency. *Ace Services, Kansas*. USEPA, 1 May 2003. http://www.epa.gov/region7/cleanup/npl_files/ksd046746731.pdf (23 September 2004).

———. *NPL Site Narrative for Ace Services*. USEPA, n.d. http://www.epa.gov/super fund/sites/npl/nar1449.htm (23 September 2004).

———. *NPL Site Narrative for Pantex Plant (USDOE)*. USEPA, n.d. http://www.epa .gov/superfund/sites/npl/nar1314.htm (24 September 2004).

———. *NPL Site Narrative for Wright Ground Water Contamination*. USEPA, n.d. http://www.epa.gov/superfund/sites/npl/nar1464.htm (24 September 2004).

———. *Pantex Plant (USDOE), Texas*. USEPA, n.d. July 2004. http://www.epa.gov/ earth1r6/6sf/pdffiles/0604060.pdf (24 September 2004).

———. *Wright Ground Water Contamination, Kansas*. USEPA, n.d. http://www.epa .gov/region7/cleanup/npl_files/ksd984985929.pdf (24 September 2004).

U.S. Geological Survey. *High Plains Aquifer Background Information*. http://www .ne.cr.usgs.gov/highplains/bckgrnd.html (29 April 2002).

White Plume, Debra. *A Concise History of Colonization Efforts Against the Oglala Lakota Oyate on Pine Ridge: With Comments Regarding Current Political and Social Conditions*. Colonialism, Resistance, and Decolonization, March 2000. http:// www.fireonprairie.org/Colonization%20Pine%20Ridge.html (4 August 2003).

INDEX

low well pressure, compensations for,
169–70
Lubbock Extension Service, 182, 258
Lubbock, Texas, 23, 54, 179–81, 218,
247–48, 257
LUST (Leaking Underground Storage
Tank) sites, 51–52
Lyle, Bill, 164

McKusick, Vincent L., 200, 208–9
mammoths, 116, 120–21, 123, 125
Mangelsdorf, Martha, 152
Manila Water Company (Philippine
Islands), 234
manure, 259
Mapp, Harry P., 31
Marsh, Alvin, 195
Mason City, Iowa, 139
Mason, John, 88–93, 103–6
Masterson, Bat, 136
Maximum Contaminant Levels, 47
Medicine Root Creek, 274
Meltzer, Dave, 119
Mesa Group, 230
Mesa Petroleum, 214–15
MESA system (mid-elevation spray
application), 165
Mesa Vista Ranch, 215, 220, 225–26
Mesa Water, Inc., 221–22, 224, 228–32,
236
Mescalero Escarpment, 1, 3, 16, 271
mesquite, 19, 34
metabolic rate of plants, 46–47
Mexico, 133, 250
microfossils, 92
Midland, Texas, 18

millet, 59, 60, 266
milo, 217, 238
mined basins, 68
Mississippi River, 250
Missouri River, 27, 42, 54–57, 134
Mni Wiconi project, 55–57, 251
monitoring wells, 223–25
Moss, Delbert, 197–98
mud pit, 96
municipal water systems:
Amarillo Water, 218, 221, 226,
228–30
Colby, Kansas, 49–51
Colorado River Municipal Water District, 252
CRMWA, 217–36, 238–40
plunging water levels, 216–17
privatization, 233–35
sources, 10, 12, 92, 189
treatment facilities, 44
water quality, 47, 50–51
wells, 28, 65, 217
Museum of the Llano Estacado, 136, 139

National Priorities List (Superfund), 49,
50, 52
Native Americans:
American Indian Movement, 272–73
aquifer draw-down levels, 67
Lakota Nation, 54–57, 77, 83–84,
272–73, 281–82
Pine Ridge Indian Reservation,
54–57, 251, 271–82
Plains tribes, 9–10, 83, 124–26
Rosebud Indian Reservation, 54, 56
Wounded Knee, 272–73

the communities," she emphasized. "That's happening now, and I think it will accelerate. There are some at Tech who are looking at returning to the grassland and grazing system. Go back to cowboys and windmills. We could be very romantic about this."

"Should we replace the cows with buffalo, because they're better suited to this landscape?"

"No."

"Some people think—"

"'Cause the song says this is where the buffalo roam." She smiled. "In some areas it's easier than it is in others. The manure is the problem. If I just put it on dry ground and leave it out there, it's going to dehydrate, and then when it rains, I've got reconstituted . . ." She let the sentence hang. "Or I could use it as a soil amendment fertilizer, if I'm growing a crop. But that's got to be managed according to agronomic requirements. It can be done, if it's done carefully. I think there are people who do a good job of it. There are some who don't. The question is how to manage the overall system. How to spread the buffalo out far and wide."

"If they actually roam," I suggested.

"If they actually roam," she agreed. "If there aren't things that inhibit them from doing that. The same thing holds with cows, by the way. There are ways to distribute their water tanks and control the traffic. Encourage them to move to the other side of the field. Rotational grazing. It's a matter of shifting the animals in a controlled pattern. They condition to it very well, because they're always moving to better grass."

Though she did not mention them by name, Porter was undoubtedly thinking of Frank and Deborah Popper as she responded to my buffalo question. The Poppers hail from New Jersey: he teaches land-use planning at Rutgers University; she is a geography professor at the Staten Island campus of the City University of New York. They have proposed what they call a "Buffalo Commons," a vast, fenceless buffalo pasture covering most of the Great Plains. Declining water tables and declining populations, they argue, demonstrate that farming is not sustainable here. "Tear down the fences, replant the shortgrass, and restock the animals, including the buffalo," they wrote in the 1987 paper in which they first proposed the idea. "The federal government's commanding task on the Plains for the next century will be to recreate the nineteenth century."

> In many areas, the distinctions between the present national parks, grasslands, grazing lands, wildlife refuges, forests, Indian lands, and their state counterparts will largely dissolve. The small cities of the Plains will amount to urban islands in a shortgrass sea. The Buffalo Commons will become the world's largest historic preservation project, the ultimate national park. Most of the Great Plains will become what all of the United States once was—a vast land mass, largely empty and unexploited.

Predictably, the concept was vilified. The Buffalo Commons, wrote then-governor Mike Hayden of Kansas, made "about as much sense as suggesting we seal off our declining urban areas and preserve them as a museum of twentieth-century architecture." (Hayden has since come around to a position of cautious support.) Angry ranchers showed up at meetings to denounce the idea and its promoters. The Poppers insisted they were misunderstood. All they were saying, Deborah Popper told the Lawrence, Kansas, *Journal-World* in early 2004, "is that when the Plan A Economy—that is, agriculture—fails, there needs to be a Plan B Economy" based on ecologically sound principals. They have called the idea of the commons a metaphor that got out of control. "Early on, we were accused of being for some sort of forced, government-led land grab," Frank Popper complains. "I have no idea where that came from."

How much sense the Buffalo Commons idea makes depends very